T5-AFN-028

THE STATE OF THE WORLD'S PARKS

Also of Interest

Issues in Wilderness Management, edited by Michael Frome

†*The National Park Service,* William C. Everhart

Managing Air Quality and Scenic Resources at National Parks and Wilderness Areas, edited by Robert D. Rowe and Lauraine G. Chestnut

The Imperial Lion: Human Dimensions of Wildlife Management in Central Africa, Stuart A. Marks

Battle for the Wilderness, Michael Frome

International Dimension of the Environmental Crisis, edited by Richard N. Barrett

†*Environmental Planning and Management,* John H. Baldwin

Managing Renewable Resources in Developing Countries, edited by Charles W. Howe

Conflict and Choice in Resource Management, John S. Dryzek

Valuation of Wildland Resource Benefits, edited by George L. Peterson and Alan Randall

†*A Wealth of Wild Species: Storehouse for Human Welfare,* Norman Myers

Natural Resource Administration: Introducing a New Methodology for Management Development, edited by C. West Churchman, Spencer H. Smith, and Albert H. Rosenthal

†Available in hardcover and paperback.

About the Book and Authors

There are currently more than one thousand national parks around the world, from Yellowstone in the United States to Dürmitor National Park in Yugoslavia. Many of these parks are threatened by poaching, development, pollution, overuse, and a host of other problems. *The State of the World's Parks* is the first systematic, international study of the pressures these special places face. The book examines two questions—what are the threats to national parks worldwide, and what can be done about them?

The authors borrow from the disciplines of ecology, anthropology, and sociology, as well as from the applied fields of wildlife management, park management, and environmental science. Their arguments are based on data collected from a survey of 135 parks in more than fifty countries; the voices of park managers, conservation officers, and game wardens are heard throughout the book. Recommendations, some controversial, are made for resource management, policy, and research. *The State of the World's Parks* is important reading for conservationists, park managers, policymakers, scientists, and citizens interested in national parks and their future.

Gary E. Machlis is associate professor of forest resources and sociology at the University of Idaho. He also is the sociology project leader for the University's Cooperative Park Studies Unit, a research station of the U.S. National Park Service. He holds a Ph.D. in human ecology from Yale University and has written widely on the sociological aspects of park management. *David L. Tichnell* has a master's degree in wildland recreation management from the University of Idaho. He currently is a Peace Corps volunteer in Santiago de Pursical, Costa Rica, helping to develop a forest nursery for the local community.

The authors gratefully acknowledge the support
of the International Affairs Office
of the U.S. National Park Service
and of the World Wildlife Fund-U.S.

THE STATE OF THE WORLD'S PARKS

An International Assessment for Resource Management, Policy, and Research

Gary E. Machlis and
David L. Tichnell

Foreword by Hal K. Eidsvik

Westview Press / Boulder and London

Published in 1985 in the United States of America by Westview Press, Inc.; Frederick A. Praeger, Publisher; 5500 Central Avenue, Boulder, Colorado 80301

Library of Congress Cataloging in Publication Data
Machlis, Gary E.
The state of the world's parks.
Bibliography: p.
Includes index.
1. National parks and reserves. 2. National parks and reserves—Management. 3. Human ecology. I. Tichnell, David L. II. Title.
SB481.N32 1985 333.78'3 85-10512
ISBN 0-8133-0274-9

Printed and bound in the United States of America

10 9 8 7 6 5 4 3 2 1

Contents

Illustrations

Figures

Tables

Foreword

National parks are a global phenomenon found in more than 120 countries. These parks are as diverse as the physical settings and cultural patterns of the many nations that have established them, yet within each country they serve as part of the cohesiveness that binds people together. Parks reflect—and help create—people's pride and love for their national heritage.

Parks are a creation of a political process. As such, they are a microcosm of the political scene and, like all societies, they are subject to such dangers as nuclear war and population explosion. Without stable societies there can be no stable parks. In the Sahel of Africa, for example, drought, civil wars, and social disruption are rampant. The parks do not fare well.

National parks are also a reflection of natural processes. When these processes are given adequate consideration in park establishment and management, it is likely that a park can remain ecologically stable. But when, for example, wild populations are squeezed into inappropriately designed areas, as sometimes happens, the result can be habitat destruction and disharmony with neighboring users as both wildlife and people fail to recognize park frontiers. *The State of the World's Parks* focuses upon threats to natural systems, describing and documenting the extent and origins of threats to parks around the world and setting forth potential solutions to these problems. The book is an important contribution that will assist the International Union for Conservation of Nature (IUCN), World Wildlife Fund, and other conservation organizations in carrying out their obligations.

The audiences for this book are as varied as the parks themselves. For professional park managers, the book identifies potential dangers to long-term effective management and shows that many parks face common problems. Understanding of potential threats should enable managers to take corrective action before major perturbations of natural values occur. Training officers will find that this book provides a wealth of information that can be used to focus their programs to ensure that managers, wardens, rangers, and field units are alert to threats and are trained to cope with them at the appropriate level of action.

For scientists, Machlis and Tichnell deliver a systematic and comparative study, based on a unique set of carefully collected data. They identify several major areas for future research. For instance, although the role of protected areas in monitoring wildlife populations has been known for years, the potential of using parks for monitoring levels of toxic chemicals in soils and ground water is less commonly understood. How serious are external threats to parks from acid precipitation, dust, smoke, and other chemicals? How can management plans be improved to negate threats to park resources?

Although not site specific, the book identifies significant threats for citizens' groups and points out the tradeoffs between preservation, conservation, and development. Citizens can determine how threats relate to protected sites within their area of interest.

For some people, it may be an embarrassment to have threats to parks so explicitly examined; for most, it should be a blessing and an invaluable tool for conservation of protected areas. The conservation process is dynamic; once a threat is known, energy and resources can be focused upon its elimination. Without knowledge there can be no cure—without a cure there may be no parks. The authors of *The State of the World's Parks* have made a major contribution to park management with a work that should help protected areas fulfill their purposes.

Hal K. Eidsvik
Senior Policy Adviser, Parks Canada
Chairman, IUCN Commission on
National Parks and Protected Areas

Preface

The First World Conference on National Parks was held in Seattle, Washington, during the summer of 1962. One of the early speeches at this historic meeting was made by M. A. Badshah, then the wildlife officer for the State of Madras, India. Midway through his presentation, the speaker provided a startling and personal vision:

> So far as I can visualize, the nearest approach to a paradise on earth—short of a fabled land with milk and honey flowing and angels and fairies in attendance—is a national park, a combination of some or all of the ingredients: forested hills and valleys, sparkling multicolored lakes, crystal clear streams, rippling brooks, placid rivers, silvery beaches, scented flowers, luscious fruits, sweet berries, snow-clad mountains, and wildlife. In the parks coexistence, tolerance, goodness, love, and attention prevail, and both man and beast can go each his own way and wander freely. (1962:27)

Unfortunately, no national park has matched the wildlife officer's vision, and it is doubtful that any will. Perhaps such joyous rhetoric is an unfair burden to place on national parks, for to satisfy such claims would defy much of what we know about Nature, society, and ourselves. Yet if national parks are a kind of earthly paradise, they are a paradise in trouble. Chemical pollution, erosion, poaching, fire, loss of habitat, overuse, and a host of other real-world problems are increasingly common to them. This book attempts to assess the problems these special places face around the world. We hope to answer two questions—what are the threats to national parks worldwide, and what can be done about them?

Several individuals and organizations have aided our effort. Robert Milne of the International Affairs Office of the U.S. National Park Service has been especially helpful, offering advice, constructive criticism, and information. He introduced us to R. Michael Wright of the World Wildlife Fund (WWF), and Wright's willingness to provide a small WWF-U.S.

research grant enabled the data collection to begin. We are grateful to the WWF, and hope our study is a useful contribution to their worldwide activities. The views expressed, however, are only our own and not those of the WWF.

The project was generously aided by the reviewers who suggested improvements to the survey, particularly Rabel Burdge, William Burch, Don Dillman, Raymond Dasmann, Edwin Krumpe, Robert Lee, Roland Wauer, R. Michael Wright, and R. Gerald Wright. In addition, James Agee, William Burch, Don Field, Milford Fletcher, Ro Wauer, and R. Gerald Wright critiqued earlier versions of the manuscript. The editing by George Savage, Jean Matthews, Susan Heib, and Pat Peterson is appreciated.

Joan Klingler has again been a full partner in our research—typing correspondence and numerous drafts, preparing and checking the statistical tables, finding lost references, and never losing patience.

The College of Forestry, Wildlife, and Range Sciences at the University of Idaho, with its emerging emphasis on international resource issues, has been a patient supporter of our efforts to learn about people and parks.

Finally, we would like to express our gratitude to those national park managers around the world who participated in our efforts to document threats to parks; theirs is a crucially important task.

Gary E. Machlis
Moscow, Idaho

David L. Tichnell
Santiago de Pursical, Costa Rica

THE STATE OF THE WORLD'S PARKS

CHAPTER 1

Introduction

National parks and reserves are an integral aspect of intelligent use of natural resources.

—John F. Kennedy

National parks are a particular and uniquely human use of Nature. Scattered across the globe, located in diverse climates, continents, and countries, they are managed by different governments and visited by peoples of many cultures. They exist in a world of competing claims for scarce resources, one of unintended consequences and natural change—both evolutionary and short term. Parks are a unique way to use Nature, but—as Kennedy perhaps implied—not the only, most preferred, or most practiced.

So, it is not surprising that national parks around the world are faced with increasing demands—to house more wildlife, to entertain more visitors, to absorb more pollutants, and to offer up more resources for agriculture, forestry, and mining. These demands are not without impacts. To use a term central to this book, national parks may be increasingly *threatened,* and it is our intention to examine in some detail the status, extent, and causes of such threats. We hope to answer two questions: What are the threats to national parks around the world, and what can be done about them?

The current debate surrounding these questions is not atypical of other international environmental controversies. Indeed, later chapters will illustrate that threats to parks are intertwined with the problems of economic development and population growth, among others. Often the issue is phrased simplistically as preservation versus use; general concepts like "ecodevelopment" or "regional planning" are sometimes proposed as solutions. Much of the rhetoric is emotional, well intentioned, and tinged with self-interest (see, for example, Frome 1981). Few of those involved in the debate must live with the day-to-day consequences of national park policies. The voices of field managers—park superintendents, wardens, rangers, conservation officers—as well as those of local citizens have not been widely heard.

Nor has the discussion been infused with systematic inquiry. Almost all scientific evidence has related to a specific country, park, or species (Nelson et al. 1978); comparative analysis of an international population of parks has not been attempted. Certainly, case studies are useful: They bring into focus key variables and provide uniquely important details. Their specificity and drama are attractive to scientists and policymakers alike. Yet it is difficult (and sometimes dangerous) to generalize from them; broad trends and predictable relationships are not so well revealed. It is these broad trends that we hope to discover.

Our approach is to examine the state of the world's parks from an interdisciplinary perspective and to base our argument on data gathered systematically from a representative sample of the world's national parks. We borrow from the academic disciplines of general ecology, human ecology, systems theory, anthropology, and sociology, as well as from the applied disciplines of wildlife management, park management, and environmental science. We are cognizant of the difficulties in merging various scientific disciplines to deal with a particular resource issue (see Klausner 1971). We are also aware of the political notoriety that threats to parks have gained in the last few years and seek to provide an empirical basis for thoughtful discussion. Therefore, some care is taken to explain the assumptions, concepts, theoretical concerns, and methodological techniques that underlie our research.

OVERVIEW

We begin by identifying four major assumptions concerning human behavior and the environment. These assumptions, although not new, stress that Nature is embedded in both biological and sociological systems and that care must be taken to unravel the relationships between people and natural resources. Next, we tentatively describe two crucial concepts that are intuitively easy to grasp but surprisingly difficult to define formally: (1) national park, and (2) threats.

Chapter 2 recounts how the problem of threats to parks has evolved. The discussion illustrates that national parks are created, sustained, and altered—in part—by the dynamic interactions between human populations, culture, and the biophysical environment. Therefore, in Chapter 3 a human ecological perspective is applied to parks to gain a better understanding of these interactions. Several biological and social variables that emerge as important from the discussion are carefully described.

Chapter 4 begins with a brief, nontechnical description of the methods used to gather and analyze the data. The results are then put forward in some detail, with numerous tables for those readers wishing to interpret the information for themselves. The material moves from general results to more complex analyses. In Chapter 5, we present the major conclusions that can be derived from the data and offer a set of specific recommendations to international organizations, national park agencies, and scientists conducting research on threats to parks. We conclude with a comment on the perils, opportunities, and choices that characterize the state of the world's parks.

ASSUMPTIONS ABOUT MAN AND NATURE

As the naturalist John Livingston has written in *The fallacy of wildlife conservation* (1982), "it is the assumptions themselves that ultimately determine the persuasiveness of our arguments." This is especially true for the management and protection of national parks, because park proponents too often have failed to articulate the articles of faith that guide their policies and proposals. Likewise, interdisciplinary research such as this can quickly become unintelligible unless major assumptions are made explicit. The four assumptions that underlie our examination of threats to parks are discussed below.

Assumption 1. *Homo sapiens* is a natural species, that is, a biological species constrained by Nature. A significant portion of human social behavior is biologically determined.

Support for this assumption comes from the emerging disciplines of sociobiology and evolutionary ecology. Sociobiology, according to E. O. Wilson (1975), is "the scientific study of the biological basis of all forms of social behavior in all kinds of organisms, including man." The discipline has emerged out of the animal ethology studies carried on since World War II and recent advances in genetics. Wilson's *Sociobiology: The new synthesis* (1975) helped define key sociobiological issues, and the entire field has stirred considerable controversy (see Sahlins 1976; Caplan 1978).

Sociobiologists argue that when social behavior is related to natural selection, an organism should behave in such a way that it maximizes the fitness of its genes—that is, its ability to reproduce. Studies of parent-child relations, sibling rivalry, and kin networks in higher primates and humans have selectively supported this idea (Barash 1982). Biological evolution has set broad constraints on *Homo sapiens'* behavior; within these constraints, culture and environment steer our course. Such con-

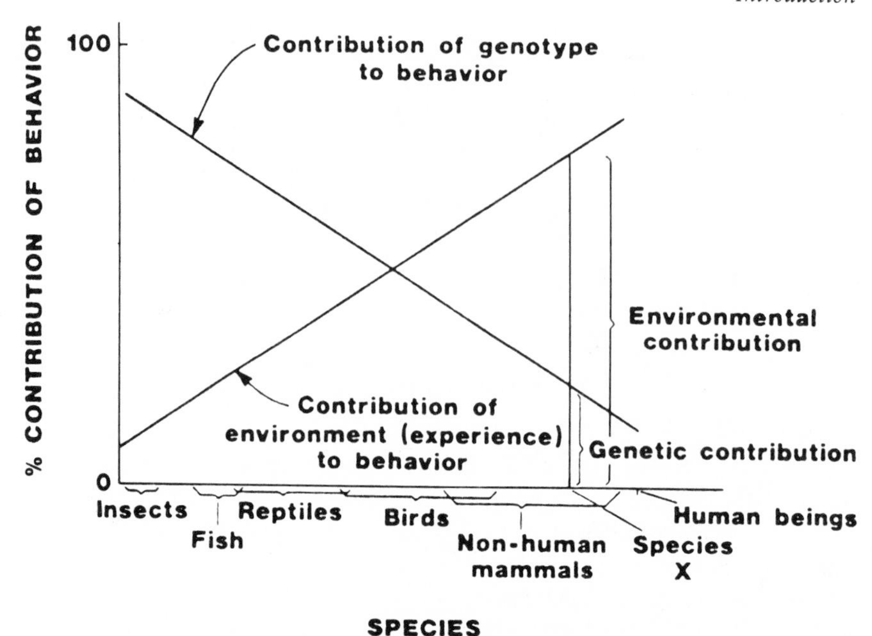

Figure 1.1. Relative contribution of genetic and environmental factors to the behavior of different animals. For any given species (X), the sum of genetic and environmental components is 100 percent. (Technically, a third component should be added: an interaction effect between genotype and environment.) Note that genetic and environmental contributions are each above zero for each species. Reprinted by permission of the publisher from "Evolution and behavior: Getting it together," chap. 3, by David P. Barash, *Sociobiology and behavior*. Copyright 1978 by Elsevier Science Publishing Co., Inc.

straints can be considered biosocial demands that must be met. Examples include nutritional minimums, the long period of infant care in humans, and the need to order sexual relations (Harris 1980; Burch and DeLuca 1984).

Natural selection theory also provides the foundation for evolutionary ecology, which focuses on explaining through selection processes the key interactions within and among populations (for a review see Emlen 1973; Pianka 1983). Topics addressed include foraging behavior (Krebs 1978), spatial organization (Davies 1978), and life history strategies (Stearns 1977). E. A. Smith (1984) reviews the application of these researches to human behavior.

Figure 1.1 illustrates that the contribution of genetically determined behavior to the human repertoire may be small relative to other species. Yet current studies of mental disorders, homosexuality, aggression, and parent-child bonding confirm the existence of important biosocial behaviors (Wilson 1978). The assumption that such behaviors exist is crucial to park resource management, for it suggests that a set of human behaviors occurs in and around parks that is predictable and universal. Such common regularities as the need for sustenance, shelter, and protection of the young may guide significant portions of the human behavior that in turn impact park resources.

Assumption 2. *Homo sapiens* is also a unique species, with special traits that make it wide ranging, dominant, and capable of wide behavioral variation.

Humans have the unique capability of complex language and can communicate across time (with the written or recorded word) and space, unlike other social animals (Chomsky 1972; Lumsden and Wilson 1981). We use symbols that transmit emotional messages as well as information and employ myth and religion to guide behavior into morally approved channels. Our technologies, even among the most nonindustrialized human populations, are elaborate and powerful. Among industrialized societies, they are characterized by large inputs of energy, resources, and information. Our organizational skills are unparalleled in Nature: They include complex hierarchies, intricate rules or norms for behavior, and the ability to assume several different social roles almost simultaneously.

Hence, wide variety can be found in certain areas of human behavior. Considering food habits, Ian Robertson points out that "Americans eat oysters but not snails. The French eat snails but not locusts. The Zulus eat locusts but not fish. The Jews eat fish but not pork. The Hindus eat pork but not beef. The Russians eat beef but not snakes. The Chinese eat snakes but not people. The Jale of New Guinea find people delicious" (1981:62). Other areas of behavior—the way we work, make love, fight, and play—are also culturally specific.

The assumption that much of behavior will vary by culture and environment is important, for it serves as a reminder that a set of human activities related to parks is widely divergent, socially influenced, and relatively fluid and changing. Human impacts upon parks are likely to emerge as dynamic and evolving; man-Nature relationships are likely to vary in form and content from culture to culture, if not from park to park.

Assumption 3. *Homo sapiens* is ecologically interdependent with the natural world.

As the anthropologist Robert Redfield (1963) noted, all societies have world views—visions outward from themselves that explain and describe their role and responsibilities. Western industrialized societies, which

Table 1.1
Differences Between Dominant Western World View and Ecological Perspective

Topic of Assumptions	Beliefs in Dominant Western Worldview	Tenets of Ecological Perspective
The nature of human beings	People are fundamentally different from all other creatures on Earth, over which they have dominion.	Although humans have exceptional characteristics (culture, technology, etc.), they remain one among many species that are interdependently involved in the global ecosystem.
Social causation	People are masters of their destiny; they can choose their goals and learn to do whatever is necessary to achieve them.	Human affairs are influenced not only by social and cultural factors, but also by intricate linkages of cause, effect, and feedback in the web of Nature; thus purposive human actions have many unintended consequences.
The context of human society	The world is vast, and thus provides unlimited opportunities for humans.	Humans live in and are dependent upon a finite biophysical environment that imposes potent physical and biological restraints on human affairs.
Constraints on human society	The history of humanity is a history of progress; for every problem there is a solution, and thus progress need never end.	Although the inventiveness of humans and the powers derived therefrom may seem for a while to extend carrying capacity limits, ecological laws cannot be repealed.

Source: W. R. Catton, Jr., and R. E. Dunlap, A new ecological paradigm for post-exuberant sociology, American Behavioral Scientist 24(1):15-47.

have a history of technological, scientific, and economic expansion, largely share a view characterized by human exceptionalism and an exuberance for exploiting Nature (White 1967; Catton and Dunlap 1980).

In contrast to this dominant Western world view is the ecological perspective, which assumes a connectedness between human values, human behavior, and the environment. This paradigm is not new, though it is gaining renewed support as the complexity of man's role on earth becomes more apparent.[1] Rene Dubos in *The wooing of earth* states:

> The ecological image of human life that is now emerging is in part a consequence of concern for environmental degradation. It has also been influenced by the development of new scientific disciplines such as cybernetics, information sciences, general systems theory, and hierarchy theory. Its more profound origin, however, is the increased awareness of the intimate interdependence between human beings and their total environment. (1980:146)

William R. Catton, Jr., and Riley E. Dunlap (1980) provide a useful comparison of the ecological and dominant Western world views (see Table 1.1). Human nature, social causation, the context of human society, and constraints on human progress all are viewed differently by those who accept the ecological perspective, as we do in this book.

This assumption of interdependence is powerful, for it demands humility—from scientists and managers, from park visitors and citizens, from our expectations of what park management can and cannot do. The eruption of volcanoes in North America, the interruption of rainfall in Africa, floods, and disease in Asia—all remind us of the difficulties in managing Nature to meet human objectives. The construction of flood control dams, bioengineering of genetic markers, and production of thermonuclear weapons equally remind us that our special abilities have enormous impacts upon the rest of Nature.

Assumption 4. The complexities of man-Nature relations can best be understood using a general systems approach.

Perhaps no concept has been incorporated into contemporary science as extensively as the concept of system (for a historical review, see Kusel 1984). E. Lazlo notes: "'System sciences' are springing up everywhere, as contemporary scientists are discovering organized wholes in many realms of investigation. System theories are applied in almost all of the natural and social sciences today, and they are coming to the forefront of the human sciences as well" (1972:13).

The term *system* has a number of meanings; in general it is "a set of interacting units with relationships among them" (Von Bertalanffy 1956). Systems are in turn composed of subsystems—those interacting units that carry out a particular, identifiable process. A general systems theory (GST), applicable to a wide range of natural and social phenomena, has been widely proposed by L. Von Bertalanffy (1968), J. G. Miller (1978), and others. Miller states the scope of GST most explicitly: "Complex structures which carry out living processes I believe can be identified at seven hierarchical levels—cell, organ, organism, group, organization, society and supranational system. My central thesis is that systems at all these levels are open systems composed of subsystems which process inputs, throughputs and outputs of various forms of matter, energy and information" (1978:1). The GST perspective has been applied to ecology (see Odum 1983). Indeed systems thinking is almost a prerequisite for ecological thinking; viewing environmental systems holistically, rather than as a collection of unrelated objects, forms the foundation of ecology (Ricklefs 1973).

A paradox is present, however: Systems understanding is crucial to the ecological sciences, yet as Gödel's Theorem suggests, no system can fully understand itself, because more components are required to analyze and understand than simply to function. Ecologists avoid the paradox by emphasizing models—simplified conceptions of real-world systems.

Both the potentials and limitations of this technique have received attention (Halfon 1975; Kitching 1983). Ecological modeling has focused on numerous kinds of environments, such as coastal zones (Johnson et al. 1980), lakes (Scavia and Robertson 1979), and forests (Botkin and Miller 1974; Borman and Likens 1979). A few models of park ecosystems have been attempted, such as one by E. DeBellevue et al. (1979) of Everglades National Park in the United States. Models that include both biological and social components have been developed, primarily based on energy flows (Odum 1971; Marks 1984).

This system assumption is important, for it allows us to examine problems in national parks by simplifying the components of park systems into models (composed of subsystems) and gathering specific data on each subsystem. As we will see in later chapters, this strategy is central to our analysis.

With these four assumptions—*Homo sapiens* is (1) a biological species with certain biosocial behaviors; (2) unique in its cultural variation; (3) part of a wider web of Nature; and (4) best understood from a systems perspective—we turn to a brief discussion of the central notions of the book—national parks and threats.

NATIONAL PARKS DEFINED

The contemporary term *national park* was described as early as 1832 by the U.S. artist-explorer George Catlin. Arguing for the establishment of Yellowstone National Park in the United States, Catlin called for "A *nation's* park, containing man and beast, in all the wild and freshness of their nature's beauty!" (1851:262). That Catlin's proposed institution has enjoyed success is obvious: Figure 1.2 shows a steady rise in the establishment of national parks and other protected areas; the number of newly established parks has accelerated worldwide since World War II. In the decade 1972 to 1982 alone, the number of parks and protected areas increased 47 percent, and the territory encompassed by these areas grew 82 percent (Miller 1984).

As the national park idea has spread throughout the world, the title *national park* has acquired a variety of meanings (for a review see Constantino 1974; Harroy 1974; Coolidge 1978; Runte 1979). Inconsistent use of the term has been seen as detrimental to gaining popular support, to comparing national parks for scientific purposes, and to encouraging proper planning strategies (Brockman and Curry-Lindahl 1962; Forster 1973). Hence, several international conventions have attempted to standardize official definitions.

The Convention Relative to the Preservation of Flora and Fauna in the Natural State (1933) and the Convention on Natural Protection and

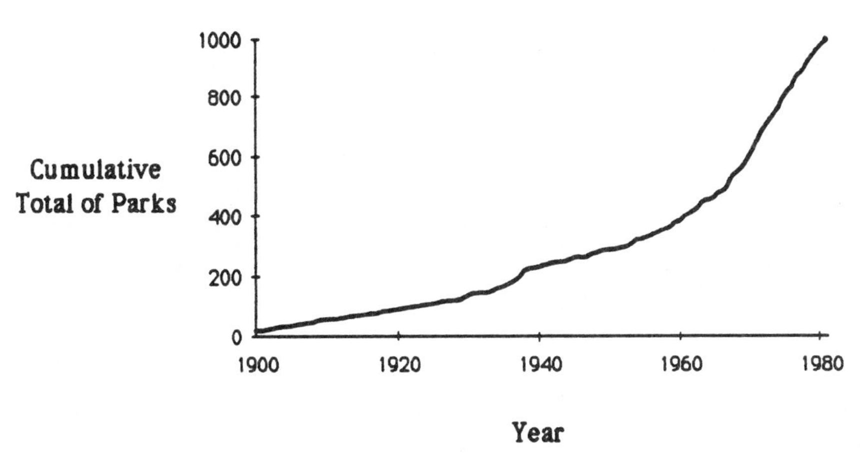

Figure 1.2. Number of national parks established worldwide, 1900–1980. From International Union for the Conservation of Nature (IUCN), *United Nations list of national parks and equivalent reserves* (Gland, Switzerland: IUCN, 1982b), p. 10.

Wildlife Preservation in the Western Hemisphere (1940) were established to encourage the development and proper management of parks in Africa and Pan America. Although they differed somewhat, the conventions generally described national parks as areas established to protect natural phenomena for public enjoyment and from human exploitation (Hart 1966).

In 1967, at the request of the United Nations (UN), the International Union for the Conservation of Nature (IUCN) published the first UN list of national parks and equivalent reserves. The list was to act as a "roll of honor" and an incentive to countries to establish parks. To be included on the list, national parks had to meet three criteria, concerning protective status, minimum size, and effective enforcement.

Increasing confusion over the term *national park* and the continuing establishment of areas that did not meet the IUCN criteria led to revisions in the definition and subsequent adoption by the Tenth General Assembly of the IUCN in New Delhi in 1969 (Forster 1973). The definition comprises the same main components as earlier definitions but is more sharply descriptive. In 1972 the UN definition was further modified to allow the inclusion of parks that may have zones where the protection of cultural heritage is the most important objective (Forster 1973). The UN definition subsequently was adopted by the Second World Conference on National Parks in 1972. Today this definition is the one most widely recognized.

These official definitions may overstate the universality of the national park concept. G. B. Wetterberg examined the motivations underlying establishment of national parks and found that the most common element is their national significance for the country in which they are found. Hence his analytic definition reads: "A portion of a nation's territory which, because of its natural and/or cultural characteristics, has been considered of such national significance that the national government of that country has legally set it apart for public benefit" (1974:27). Yet Wetterberg makes clear that disagreement still existed concerning the definition of national parks. He notes: "As it is used globally today, the term 'national park' gives very little indication about the appropriate types of management and uses to be followed in areas bearing that name" (1974:33).

Given the variation inherent in any kind of global land use classification, we have adopted the IUCN definition, which reads:

> A national park is a relatively large area:
> 1) where one or several ecosystems are not materially altered by human exploitation and occupation, where plant and animal species, geomorphological sites and habitats are of special scientific, educative and recreative interest or which contains a natural landscape of great beauty; and
> 2) where the highest competent authority of the country has taken steps to prevent or eliminate as soon as possible exploitation or occupation in the whole area and to enforce effectively the respect of ecological, geomorphological or aesthetic features which have led to its establishment; and
> 3) where visitors are allowed to enter, under special conditions, for inspirational, educative, cultural and recreative purposes. (1975: Preamble)

Our study focuses on national parks as just defined, though obviously other forms of protected areas—reserves, recreation areas, state and provincial parks—also are important units of conservation.[2] We have limited the scope of our inquiry but are aware that the problems of conservation (and perhaps their solutions) do not respect such boundaries.

THE CONCEPT OF THREATS

The concept of threats is even more difficult to define. As J. S. Burgess and E. Woolmington (1981) note, the term *threat* is highly anthropocentric: It is a social metaphor applied to biological systems. It suggests human concern that certain valued characteristics of Nature are in danger of degradation or destruction. Ecologically, threats can be considered "suspected stresses," and "thus *threat* can be said to roughly equal perceived (and sometimes imagined) *stress,* often with additional connotations relating to the interests of the perceiver" (Burgess and Woolmington 1981:419).

Various definitions of ecosystem stress have been proposed, and most emphasize that natural systems operate within a dynamic range of tolerance, making stress a relative characteristic (Selye 1952; Barrett et al. 1976; Bayne 1975). Some levels and kinds of stress, such as volcanic eruptions, natural floods, and wildfire, are natural occurrences (deFreitas and Woolmington 1980); others are man-caused.[3] Both lethal and non-lethal forms of stress can be found. Empirical studies of pesticides, fire, and radiation have shown the complexity of stress phenomena (Barrett and Rosenberg 1981); for example, Table 1.2 illustrates the significance of major stresses on the Great Lakes, and the variations in the level of stress by biotic component.

In addition, stresses to individuals may lead to population and community change (Auerbach 1981). D. Mueller-Dombois et al. (1983) list several "evolutionary stresses"—periodic fires, floods, storms, frosts, droughts, and insect pest outbreaks. Where species groups or biological communities have been exposed repeatedly to such stresses, they have probably become adapted to them. R. S. De Santo describes such a process:

> This principle may be called the *ripple effect,* suggesting that if a central factor is present that stresses a biologic system, the impacts of that one stress will have repercussions radiating outward and disturbing the system, usually in diminishing magnitude, from the simple central impact. . . . If an ecosystem is stressed, it too is modified in accommodation to that stress. (1978:15)

Finally, not all stressors are harmful to long-range viability—some (such as human vaccines) may evoke highly valued responses (Odum et al. 1979).

Hence, threats to parks are really stresses perceived to have detrimental impacts upon valued components of park ecosystems. Such a definition is almost entirely social rather than biological. To be a threat, the stress must be perceived by ourselves or with the aid of scientific instruments. Threats can range from merely suspected to fully documented, and the level of acceptable documentation is a subjective criterion.

The questions of what is a detrimental impact and what is a valued component of an ecosystem also call for nonscientific judgments (Klausner 1971; Rapport and Regier 1980). Drought may be considered a threat to socially important species like the elephant; at the same time its impact upon soil microorganisms may be left unconsidered by scientist, manager, policymaker, and citizen. Both may be stresses; one may be considered a critical stress, or threat.

Table 1.2
Significance of Major Stresses on Great Lakes Biota*

Major Stress Classes and More Specific Stress Types	Biotic Components of Ecosystems								
	Phyto-plankton	Zoo-plankton	Macro-benthos	Salmonines, Coregonines[a]	Percids, Ictalurids[b]	Esocids, Centrarchids[c]	Littoral Macrophytes	Waterfowl	Wetland Mammals
Natural stresses									
Seasonal variations	1	2							
Major storms						2	1	2	1
Harvesting									
Mammals							2		1
Waterfowl							2	1	
Fish		2	2	1	1	1			
Chemical loadings									
Nutrients	1	1	1		2	1	1	2	
Toxins[d]			1	1	1	1		1	
Physical restructuring									
Stream alteration				1	1	2	2		
Shoreworks			2		2	1	1	1	1
Power plants				2	1	2			
Exotic introductions									
Pelagic fish	2			1	1				
Sea lamprey		2		1	2				

*The number 1 implies a strong direct effect; 2 implies a significant indirect effect.

[a]Deepwater fish
[b]Midwater fish
[c]Nearshore fish
[d]Through impact on what is marketable

Source: D. J. Rapport and H. A. Regier. An ecological approach to environmental information, Ambio 9, no. 1(1980):24.

Even calling a resource problem *critical* is a qualitative judgment of how much biophysical or social change is acceptable, and it is a threshold measure (Machlis and Wright 1984). Although quantitative data may be used to estimate biophysical or social change, the setting of threshold levels of acceptable change is a qualitative decision, often based on conventional scientific wisdom, legal requirements, and so forth. Examples include determining carrying capacity of rangeland, maximum automobile emissions, and risk associated with nuclear power plants (Burch and DeLuca 1984).

Beyond a chosen threshold of acceptability, the seriousness of a problem may vary significantly. The term *seriousness* may reflect the resiliency with which an ecosystem can respond to change. A threat (such as trail overuse) may emerge that impacts a particular subsystem, but the problem may have very localized effects or last only a short period. Other problems (such as acid rain) may have long-lasting effects or be generalized throughout a park. The choice of criteria is also subjective.

As the definition of threats to parks is hopelessly anthropocentric, it is not surprising that the perception of threats has evolved with time. Early efforts at Yellowstone National Park in the United States (feeding hay to elk in winter; killing predators) were aimed at mitigating threats caused by Nature (Stottlemyer 1981). Current park practices consider some natural stresses permissible; hence they are no longer perceived as threats (Wauer and Supernaugh 1983).

This change suggests that management objectives must be considered as a component in any definition of threats to parks. Loss of a highly visible and symbolic animal population (such as the bald eagle in the United States or the panda in China) might cause less than catastrophic harm to the viability of an entire ecosystem but so jeopardize park objectives that it constitutes a clear and present threat. Also, the IUCN definition of national parks makes clear that certain forms of exploitation will necessarily be threats. Therefore, as used in this study the term *threat* denotes

> *those activities of either human or natural origin that cause significant damage to park resources, or are in serious conflict with the objectives of park administration and management.*

Numerous authors have suggested that threats originating within (Siehl 1971; Forster 1973; Sax 1980a) and outside (Kusler 1974; NPCA 1979; Garratt 1984) national parks are modifying them to a significant degree. Recent studies, particularly of African park ecosystems, support these claims. For example, deforestation from elephant browsing in

Ruaha National Park, Tanzania, has been extensively studied (Savidge 1968; Barnes 1982, 1983). Poaching has had a documented impact in several African national parks (Eltringham and Woodford 1973; Hillman and Martin 1979; Borner 1981; Marks 1984).

Scientific evidence illustrates the complexity of the issue. For example, elephant overpopulation in the 1960s in Ruwenzori National Park, Uganda, led to severe habitat deterioration, and culling was proposed as a management response. Data collected between 1973 and 1976 showed a 26 percent decline in the elephant population, largely as a result of poaching (Eltringham and Malpas 1980). These authors suggest that culling is no longer suitable. In addition, other research has shown that lower rainfall levels and overgrazing by goats also contribute to the park's range deterioration (Musoke 1980; Yoaciel 1981). Clearly, threats are dynamic phenomena.

NOTES

1. This way of thinking has impacted both science and philosophy. An example is the rise of the "deep ecology" movement, which argues that Nature has important, intrinsic values apart from any utilitarian benefits to man. For a description of deep ecology, see Naess 1973, 1984; Devall 1980; and Tobias 1984.

2. The IUCN (1984) suggests ten categories ranging from scientific reserves to multiple-use management areas. National parks are classified as Category II.

3. This explanation skirts the problem of first causes—arguments can be made that "natural" floods are often caused by inappropriate upstream human land uses, and counterarguments made that *Homo sapiens'* abilities to build and develop are dependent upon Nature for energy and so *ad infinitum*. We shall take up this problem again in Chapter 3.

CHAPTER 2

The Origins of Threats to Parks

For centuries, humans have altered the natural landscape (Wagner 1971; Tobey 1973; Jellicoe and Jellicoe 1975). In the last century, population growth coupled with technological advances magnified *Homo sapiens'* ability to change the natural environment (Brown 1981; Eckholm 1982). Today, there is little reason a priori to believe national parks are immune to these influences. Hence, global trends that result in increasing ecological change may be important sources of threats to parks.

At the beginning of the 1980s, the world's population was estimated to be 4.3 billion persons; the population is predicted to increase by 55 percent from 1975 to the year 2000 (U.S. Council on Environmental Quality 1980). The reasons for such growth are complex (see Murdoch 1980 for a thorough discussion) and center around the demographic transition occurring in developing nations. M. P. Todaro writes: "Population growth today is primarily the result of a rapid transition from a long historical era characterized by high birth and death rates to one in which death rates have fallen sharply while birth rates, especially in developing countries, have not yet fallen much from their historic high levels" (1981:160). Figure 2.1 illustrates the disparity in growth rates between developing and developed nations for the recent past, the present, and an estimated future.

As a result of this growing population, increasing pressure has been placed on biological systems in efforts to maintain or even raise per capita consumption levels of food, water, energy, and other natural resources (Holdgate et al. 1982; Brown et al. 1984). This struggle has not always been successful; Todaro (1981) reports a decline in per capita food production in Japan, Africa, West Asia, and East Asia between 1975 and 1979. The environmental impacts are significant: Arable land lost to urbanization amounted to 50,000–200,000 square kilometers (sq

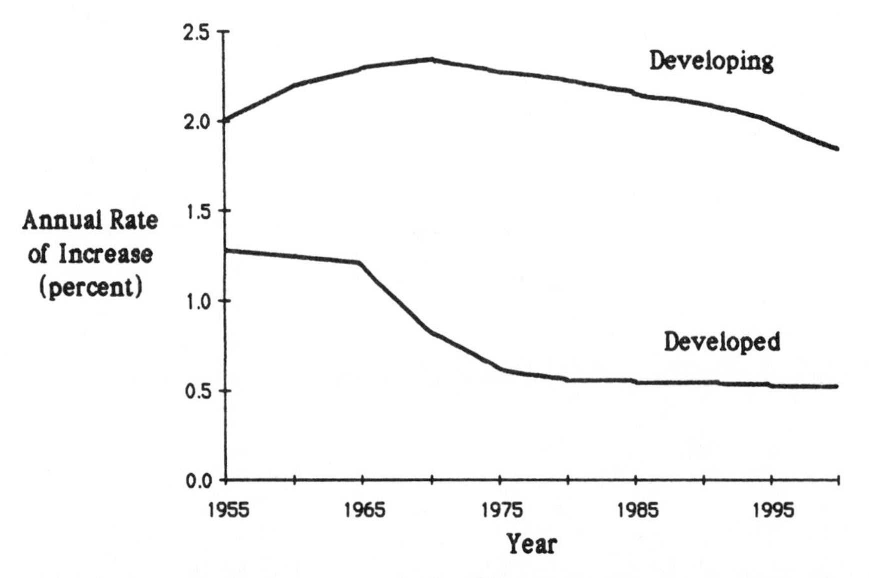

Figure 2.1. Population growth rates in developed and developing countries, 1955–2000. From United Nations, *World population trends and prospects, 1950–2000* (New York: United Nations, 1979), p. 164.

km) during the 1970s; consumption of nitrogenous fertilizers increased from 33 million to 54 million metric tons (t) and is predicted to rise to 84 million t in 1985 (Holdgate et al. 1982). In Central America, the period 1961–1978 witnessed a 39 percent decline in forested land, and deforestation in all of the tropics is predicted at a rate of 3.8 million hectares (ha) per year in the period 1981–1985 (FAO 1982).

Some peoples of the world are more severely affected by these global trends than others. Ninety percent of the population growth between 1975 and 2000 is expected in less-developed regions (U.S. Council on Environmental Quality 1980). Traditional cultures in rural areas often comprise major proportions of these societies. Poor socioeconomic conditions usually are prevalent. In contrast to industrialized societies where resource consumption is dispersed throughout a number of ecosystems worldwide, these traditional cultures often have occupied an ecological niche in local ecosystems for a long time (Dasmann 1975). Their survival has depended on a sustainable relationship with the local environment; hence their practices frequently result in the conservation of certain resources such as wildlife (Bennett 1976; UNESCO et al. 1978; Kwapena 1984).

The intrusion of more industrialized cultures has often modified this ecological equilibrium; for example, improved health care may accelerate population growth (Murdoch 1980). When coupled with traditional subsistence practices such as pastoralism and shifting agriculture, this growth can result in serious impacts on the local environment. Today, a majority of the world's rural peoples live in these modified traditional cultures, which occupy approximately one-fourth of the earth's land (U.S. Council on Environmental Quality 1980).

Regional variations in resource use, population growth, and consumption may have important implications for parks. In countries like the United States, national parks were first established on what were often perceived at the time as "worthless lands" (Runte 1979). However, national parks increasingly are being established in developing countries (Harrison et al. 1982), where they directly compete with other, often crucial, land uses (Marks 1984). This difference suggests that the "cultural reception" that the national park concept receives in developing areas may differ from responses in industrialized countries. It becomes useful, therefore, to explore how effectively the park idea has been adapted to diverse cultural situations.

THE CULTURAL ADAPTATION OF THE NATIONAL PARK IDEA

A large and growing body of literature focuses on the history of the world park movement (see Harroy 1974; Nicholson 1974; Coolidge 1978; Curry-Lindahl 1974b; Boardman 1981). Regional and national histories are available for Africa (Curry-Lindahl 1974a), South America (Wetterberg 1974), and the United States (Ise 1961; Runte 1979). Almost all stress the emergence of political institutions and management organizations, along with the evolution of the national park idea. Few have dealt with the historical *consequences* of park establishment or the history of threats to parks. However, adopting the traditional park concept has been more difficult for some cultures, particularly those in developing regions of the world such as Africa (Myers 1972; Dasmann 1975; Shelton 1983).

When the first national parks were established in developing countries, like their U.S. counterparts they were often located in remote areas, received few visitors, and thus were relatively little influenced by humans. Significant increases in park acreage and population and decreases in resources have changed this scenario. In contrast to the life-style in industrialized regions of the world, which includes a higher per capita income and standard of living, the short-term subsistence needs of indigenous people in developing regions are today acute. Hence, political and public support for parks in these areas often depends more on their contribution to local socioeconomic development than it does in developed regions (Curry-Lindahl 1974b; Myers 1981; Blower 1984).

Park management strategies have sometimes adversely affected local development. The displacement of indigenous people from areas to be established as parks, for example, has resulted in serious social impacts such as poorer nutrition and greater energy expenditure in obtaining fuel and food (Turnbull 1972; Gomm 1974; Mugangu-Trinto 1983). Africa is an instructive case. N. Myers (1984) and S. A. Marks (1984) both point out the severity of competition between African wildlife and native peoples for scarce resources. Protected park wildlife have entered villages adjacent to some parks, damaging crops and destroying livestock (Myers 1972). In addition, the dependence of local people upon local resources heightens resource demands. T. A. Afolayan (1980) estimates that 50 percent of the population in Africa south of the Sahara depends on wildlife for a source of protein.

As a partial result, popular support for parks has in some instances been more difficult to obtain. A recent survey by the African Wildlife Foundation found that a sample of Tanzanian school children think that too much land is used for parks and reserves and that these areas exist principally for the benefit of foreign tourists (Abrahamson 1983). Threats such as poaching, vegetation removal, and wildfire have increased (Curry-Lindahl 1974b; Lucas 1982; Ayensu 1984). In addition, the movement of wildlife has increasingly been restricted to park areas, which in turn have become overpopulated and have suffered ecological damage (Myers 1972).

Although problems in transposing traditional management practices to parks in developing regions have become apparent, new strategies have also evolved. Parks benefit local populations in such ways as protecting watersheds and providing opportunities for public education, but these benefits often are inconspicuous and not directly linked to socioeconomic development. Moreover, the tourism that parks attract can provide employment to local people and has become an important economic rationale for parks. It is Kenya's largest source of foreign exchange (Lusigi 1981).

Increasing tourism is not, however, a viable strategy for all parks; only certain areas have the infrastructure—the transportation facilities, hotels, and outstanding natural resources—to sustain the industry. In some instances, benefits from tourism may be evident on the national rather than the local level, and yet local cultural and economic patterns may be changed (Budowski 1977; Jefferies 1982; Olwig 1983; Machlis and Burch 1983). Consequently, local people, unaware of or uncaring about the national benefits, may not understand why areas they traditionally utilized have been set aside for visitors while their own hardships increase (Dasmann 1975; Hughes 1979; Blower 1984).

Additional, often site-specific strategies are being designed and used to help the local population adapt to the changes brought about by park management. In Kenya, for example, members of the Masai tribe, who live adjacent to Amboseli National Park, receive annual monetary compensation from the park, are furnished piped-in water for their cattle, and participate on park advisory committees (Western 1982). In Nepal, villagers around Royal Chitwan National Park legally enter that park once a year to collect grass. The presence of 63 villages within Nepal's Sagarmatha National Park illustrates how indigenous people can be permitted to remain in some parks (Mishra 1982; Shelton 1983). Other strategies of adaptation include cropping park wildlife and selling it to local people (Myers 1972) and training and employing indigenous people to work in parks and tourism enterprises (Hill 1983). Similar efforts at integrating local peoples into park management have been described for such countries as Finland (Borg 1977) and Papua New Guinea (Gorio 1978).

Despite these efforts, indigenous peoples in the rural areas of developing countries, who can potentially affect parks in significant ways, often do not perceive direct benefits from nearby parks. Perhaps the most dramatic analysis of the problem has been conducted by S. A. Marks (1984): Using detailed data on local resource use, he examined the interaction of wildlife, park management, and peoples in the Luangwa Valley of Zambia. He characterizes some current traditional techniques as "anachronistic" in contemporary Africa and pessimistically states: "It is my contention that for the West to persist in its support of preservationistic policies that hold vast acreages of land hostage to its myths is to ensure their certain destruction through African needs and perspectives" (1984:6). All ecologists, park consultants, and park professionals do not share Marks's critical view, and there are numerous examples of effective management programs, often supported by international conservation efforts (see McNeely and Miller 1984). Yet even as the national park concept is better adapted, threats may still occur as the result of a historical lack of ecological management.

ECOLOGICAL MANAGEMENT IN NATIONAL PARKS

Until recently, the primary aim of establishing parks has been to preserve specific species, special natural phenomena, or curiosities rather than ecological processes (Curry-Lindahl 1974a; White and Bratton 1980; Stottlemyer 1981; Ovington 1984). Parks have been established to protect gorillas, elephants, trees, and geological features (Ise 1961; Runte 1979). When the first national parks were established, it was apparent these features would be susceptible to threats such as wildfire and hunting. However, the significance of ecological processes in sustaining park features was not routinely acknowledged in management strategies until almost a century later (Curry-Lindahl 1972; Nelson 1978; Stottlemyer 1981).

Instead park environments were treated as static, self-maintaining entities and were protected from unwanted human and natural influences (Curry-Lindahl 1972; White and Bratton 1980). Management practices such as predator control and complete fire suppression were not altered in U.S. parks until the 1930s and the 1950s respectively (Ise 1961; Runte 1979). In some instances, protection itself was a threat: The suppression of wildfire often resulted in unnatural buildups of fuelwood in climax forests, and predator control was a factor in several parks that became overpopulated with wildlife (Stottlemyer 1981). Although studies conducted as early as the 1930s in the United States revealed the importance of unaltered ecological processes, traditional views prevailed (Wright et al. 1933; Curry-Lindahl 1974b). For example, a systematic, agencywide program concerning the management of fire was not initiated in U.S. parks until 1968 (Agee 1983). Hence, many undesirable ecological conditions continued to develop in parks.

Park management in the 1960s was characterized by a slowly growing ecological consciousness. During this period, for instance, an institute was established at Serengeti National Park in Tanzania to conduct ecological research, and several committees in the United States began to study the role of ecology in park management (Curry-Lindahl 1972; Wauer and Supernaugh 1983). Many factors contributed to this new awareness: (1) expanding scientific knowledge of ecology, (2) ever-higher numbers of obvious impacts, and (3) increasing management experience. One result was that managers began to consider managing park environments to preserve ecological conditions that existed before or when the parklands were set aside. Scientific interest in preserving a diversity of the world's ecosystems and species also grew during this period.

Now, preservation of diversity and natural change have become major objectives of the protected areas movement in contrast to the earlier preservation of disparate and static features (Harrison et al. 1982). The establishment of protected areas throughout the world has increased the diversity of reserves. However, allowing natural processes to operate freely in parks has been more difficult. Few examples have been put into practice—for the most part park managers still control fires, remove "problem" bears, and, if humans are part of the system, thoroughly regulate human activities. A dichotomy in park policy exists today between the preservation of ecological processes to maintain habitats and the perpetuation of unaltered natural change (Lamprey 1974; White and Bratton 1980). Determining which policy is most appropriate for a given park is expected increasingly to pose philosophical issues (Bratton and White 1981).

Both the public and park managers have sometimes been slow to adopt an ecological strategy, particularly if it conflicts with the protection of species; this hesitation has often slowed the mitigation and prevention of threats to parks. For example, a plan to eradicate feral burros at Grand Canyon National Park resulted in public controversy and consequently delayed implementation of the most ecologically suitable strategy. The U.S. National Park Service was forced to complete research studies to support its position before it could take action (Garrett 1978). In some African parks, elephant herds have overpopulated park lands and have caused serious ecological damage; yet park managers have occasionally been reluctant to crop them (Myers 1972). Other threats (discussed in the following sections) continue to develop from conditions perpetuated by the early lack of ecological management.

Artificial Boundaries

Parks continue to be designated by boundaries that fail to encompass the habitat required to sustain complete faunal assemblages. Few tracts of land are currently available that encompass entire ecosystems. As a result, most parks are artificial biotic areas inextricably linked to natural and human activities outside their boundaries (Curry-Lindahl 1972; Garratt 1984). The degree of this dependency may vary from ecosystem to ecosystem; for example, because of seasonal wildlife migrations East African parks are often more dependent on adjacent lands than are many North American parks (Myers 1972; Western 1982). Regardless of the degree, however, threats originating from incompatible activities on adjacent lands are among the most serious problems facing park managers today (NPCA 1979; USDI, NPS 1980).

Creating boundaries encompassing enough land to protect complete or nearly complete ecosystems is difficult to achieve, both in theory and in practice (Nelson 1978). For example, boundaries in marine reserves are especially difficult to establish (Wolff 1982). And as N. Myers and D. Myers (1982) point out, environmental problems such as industrial air pollutants that originate far from their eventual deposition are becoming more prevalent. In one study, T. V. Armentano and O. L. Loucks (1983) examined air pollutants from industrial sources in the U.S. Great Lakes region, and found that they significantly threatened several national park ecosystems. The authors note: "Pollutant concentrations in the southern sites . . . are above the thresholds that are known to produce stress, foliage damage, and altered growth rates on many sensitive species, including the coniferous trees" (1983:311). However, even if a park does have sufficient boundaries, threats often still occur because of activities within them.

Preservation Versus Use

For decades increasing numbers of visitors and subsequent efforts to accommodate them have been a source of ecological impacts in national parks (Runte 1979). The conflict originates from the paradoxical mandate that parks provide opportunities for public enjoyment in environments that are to be preserved. This mandate has been perpetuated by many mechanisms and strategies, such as the 1933 London Convention, developed to assist and encourage nations to establish parks (Coolidge 1978). A lack of policy guidelines to balance use and preservation has established the conflict worldwide (Forster 1973).

The greatest impacts often have occurred in countries such as Japan and the United States where the population has more leisure time, greater per capita income, and efficient transportation. Parks in these countries generally receive more visitors; the resources and expertise are available to develop tourism infrastructures, such as hotels and restaurants. These in turn increase the likelihood of human impacts in parks (Forster 1973; Clawson 1974). However, overuse may impact even the more isolated parks, as the instance of the Galapagos National Park, Ecuador, demonstrates (deGroot 1983).

In industrialized countries such as the United States, tourism was promoted in parks from the beginning by those seeking support for the park concept and by entrepreneurs seeking profits (Runte 1979). As the number of visitors increased, particularly after World War II, the development of accommodations and other services expanded. Although the adverse impacts of this policy had been documented much earlier, they did not become major management issues until the 1960s (Lee 1968; Runte 1979; Lemmons and Stout 1982). By 1982, visits to U.S. national parks exceeded 300 million annually (Dickenson 1984).

During the late 1970s, efforts were begun to minimize future development and to protect the natural integrity of parks. However, the number of visits has continued to grow, as has the conflict between preservation and use. In addition, the problem has begun to recur in developing countries where governments justify parks primarily on the economics of tourism (Harroy 1974).

Homo Sapiens as a Park Inhabitant

Although *Homo sapiens,* the visitor, has been a concern for decades, the ecological role of traditional, indigenous peoples has only recently been acknowledged (Mishra 1982). Past policies normally have required the removal of such peoples from park lands (Dasmann 1975; di Castri and Robertson 1982). Since indigenous cultures often occupy an important ecological niche (Dasmann 1975, 1984), their sudden, induced absence can cause undesirable natural changes. Thus, indigenous practices sometimes are artificially imitated by managers in an attempt to sustain the desired conditions. For example, R. Gomm (1974) suggests that an antipoaching campaign and subsequent removal of indigenous hunting peoples in a Kenyan park resulted in overpopulation of elephants, whose numbers eventually had to be cropped. In short, the ecological desirability of removing indigenous peoples from park lands has increasingly come into question (Myers 1972; Nelson 1978). The noted ecologist Raymond Dasmann reflects:

> Most of the land we designate as formal wilderness or set aside in national parks is land passed on to us by people who considered it to be, in part at least, their homeland. We consider it to be of national park quality because they did not treat it the way we have treated land. Too often they have gone, and our legal designations, our wardens and patrols, take their place. Something seems to have gone wrong, somewhere along the way. (1984:668)

THE LITERATURE ON THREATS

As these serious pressures on park resources have increased, research and discussion concerning threats also have grown. The literature on threats is fragmented, ranging from general treatments in popular articles (Frome 1981; Wolf 1982; McCloskey 1984) to more technical proceedings from meetings, conferences, and congressional hearings (Elliott 1974; U.S. Congress 1976).

Although a few comparative studies among parks in different countries of the world have been conducted (Goddard 1961; Hart 1966; Wielgolaski 1971; Nelson et al. 1978; Gardner and Nelson 1981), most inquiries have been limited to general case studies of specific parks (for example, Olwig 1980; Jefferies 1982; Mishra 1982) or discussions of geographically homogeneous parks (Darling and Eichhorn 1967; Myers 1972; Sax 1980b; Lusigi 1981). In addition, much of the research has been nonquantitative. Recently, a few systematic studies of threats have been conducted, most of which deal with U.S. parks.

Through mail surveys, interviews, workshops, and on-site investigations, W. E. Shands (1979) examined the relationship between federal and adjacent lands. Personnel from the thirty-five U.S. national parks included in the study were asked about potential resource impacts. Although impacts were not clearly defined, and most of the presented data were anecdotal, Shands concluded that federal lands in the United

States were affected significantly by activities occurring outside their boundaries. J. L. Sax (1980a) found that privately owned lands within U.S. parks present a substantial threat as well. Other studies (Kusler 1974; NPCA 1979) have also concluded that external pressures are a serious problem. A National Parks and Conservation Association (NPCA) study surveyed U.S. park superintendents and used some data previously collected by Shands. The analysis was based on the number of park units reporting a particular phenomenon, and the results illustrate the widespread nature of external threats: "Nearly two-thirds of the 203 respondents stated that their units suffered from a wide variety of incompatible activities on adjacent lands that affect the parks in every conceivable manner—everything from trespassing livestock that trample vegetation to industrial dyes that regularly change the color of a park creek" (1979:4).

At the request of the U.S. Congress, U.S. National Park Service conducted in 1980 its first servicewide study of threats to U.S. parks. A survey questionnaire was sent to all 326 units in the system. In the first part of the questionnaire potential threats were listed under seven major threat categories, such as air pollution and aesthetic degradation. Respondents were requested to indicate how adequately each threat was documented, whether the source of the threat was internal or external to the park, and if the threat was addressed in the park resource management plan. If a threat was reported, respondents were further asked to list its specific sources and the resources that were affected.

The study concluded that none of the parks was immune to threats. Specific findings included the following: The sources of more than 50 percent of the reported threats were external to parks; threats located within park boundaries were significantly impacting park resources; and 75 percent of the reported threats were inadequately documented. In addition, parks included in the Man and the Biosphere Programme (which emphasizes monitoring) reported threats three times more frequently than other parks (see Table 2.1). Although the results have several limitations, the study provides empirical evidence that managers of U.S. national parks have perceived significant threats to the resources under their care.

More recently, the state of California conducted a comprehensive analysis of the entire state park system (State of California 1983). Data on threats to air, water, soil, vegetation, and animal life were collected for all 267 units of the park system. Managers were asked to judge the severity of each threat and their sense of priority in dealing with it. These ratings were combined into a series of scores; Table 2.2 presents the 12 most critical threats based on these scores. Interestingly, 8 of the 12 relate to administration and management.

Table 2.1
Average Number of Threats Reported for Selected Classification of Park Areas

Type of National Park Unit	No. of Areas Reporting	Average No. of Threats
Cultural areas	126	11
Natural areas greater than 10,000 acres (4,047 ha)	85	23
Natural areas greater than 30,000 acres (12,141 ha)	63	24.5
NPS international biosphere reserves	12	36.3
Servicewide total	--	13.6

Source: U.S. Department of the Interior, National Park Service, State of the parks 1980--a report to Congress (Washington, D.C.: Government Printing Office, 1980), p. 18.

Table 2.2
Twelve Most Critical Threats to the Natural and Scenic Features of the State Park System

Potential Threat	Total Units and Projects Where Known and Suspected Threats Reported	Combined Total of Severity and Priority Ratings
Sense of unsupportive atmosphere	113	5.37
Lack of actual staff, funds	226	5.21
Beach sand erosion, loss	87	5.17
Conflicting political claims and desires	207	5.14
Plant damage by wildlife	149	5.10
Poor management of adjacent land	174	4.91
Inadequate resource policy and planning materials	156	4.65
Inholdings in our properties	94	4.64
Inadequate, inappropriate facility development	153	4.63
Unsafe facilities, features	163	4.63
Wildfire destruction of organic soils	106	4.62
Wildfire exclusion: understory buildup, changed vegetative composition	89	4.62

Source: State of California, Stewardship--1983: Managing the natural and scenic resources of the California State Park System, unpublished report (Sacramento: California Department of Parks and Recreation, 1983).

The *IBP survey of conservation sites: An experimental study* (Clapham 1980) represents a major international research effort to inventory conservation areas. The purpose of the study was to collect ecological information about conservation sites worldwide and to experiment with the survey methodology. One open-ended question dealt with human impacts upon the site. Questionnaires in seven languages were sent to sites all over the world (including some national parks), though the distribution was not systematic and the questionnaire varied from country to country. Of the 13,000 checklists distributed, 3,010 were returned. Based on the 1973 IUCN list of national parks and equivalent reserves, 22 percent of the world's national parks returned surveys; however, 80 percent of these were from temperate regions. Although detailed response rates are presented, the data on threats unfortunately are not summarized. A. R. Clapham does make several general comments such as "fire was recorded as an impact in many sites" (1980:328).

More recently, the IUCN prepared a list of the world's most threatened protected areas (IUCN 1984) to increase public awareness and support for conservation activities in these areas. Information was solicited from park professionals, scientists, and other experts, and 43 sites were initially chosen. Table 2.3 summarizes the kinds of threats reported—inadequate management resources and human encroachment were the most common. From the initial group, 12 sites were chosen on the basis of documentation, conservation value, severity of threat, need for international support, and other factors. The list includes such areas as Garamba National Park in Zaire, Mt. Apo National Park in the Philippines, Manu National Park in Peru, and Yugoslavia's Dürmitor National Park. The report from Mt. Apo National Park provides a typical example:

> The park has suffered in recent years as a result of illegal logging, encroachment by shifting cultivators and the establishment of permanent squatter settlements. In addition, the Philippine Government is reclassifying 32,000 ha of illegally cleared land within the present park boundaries for agricultural development. The park is not under an effective management regime and few regulations are enforced. Another complication is the presence of insurgents in the park. A recent drought and uncontrolled fire have also been detrimental to 30 percent of the park. It is estimated that only half of the park is still viable as a natural resource. (IUCN 1984:6)

The IUCN list is actually a compilation of case studies, and its usefulness lies in the politics of conservation rather than analysis of threats to parks. Likewise, A. R. Clapham clarifies the limitations of their experiment and offers numerous suggestions regarding similar

Table 2.3
Summary of Threat Categories by Realm

Threat Category	Realm (No. of Areas Reporting Threat) Afrotropical (13 areas)	Indomalayan (8 areas)	Australian (1 area)	Neotropical (13 areas)	Nearctic (5 areas)	Palaearctic (3 areas)	Total 43
Inadequate management resources	6	7	--	3	--	--	16
Human encroachment	4	5	--	4	--	--	13
Change in water regime or hydro development	2	4	--	3	1	2	12
Poaching	5	1	--	4	--	--	10
Adjacent land development	3	1	--	4	2	--	10
Inappropriate internal development (e.g., roads)	1	2	1	4	--	1	9
Mining/prospecting	4	1	--	4	--	--	9
Livestock conflicts	4	1	--	4	--	--	9
Military activities	5	2	--	1	--	--	8
Forestry activities	3	3	--	1	--	--	7
Acid deposition/pollution	--	--	--	--	2	2	4
Degazettement/downgrading	--	2	--	--	2	--	4
Exotic species invasions	--	--	--	1	--	--	1
Total	37	29	1	33	7	5	110

Source: Commission on National Parks and Protected Areas, *Threatened protected areas of the world* (Morges, Switzerland: International Union for the Conservation of Nature, 1984), p. 2a.

surveys. He argues for improved institutional support, better designed questionnaires, more use of close-ended questions, clear definition of terms, and systematic sampling of representative ecological areas. Others have called for systematic monitoring and inventory (Polunin and Eidsvik 1979) or for a kind of "biome-ecosystem accounting" (Rapport and Regier 1980). Writing generally about environmental monitoring, D. J. Rapport and H. A. Regier stress the need for objective analysis as a precursor to management:

> The question of whether such [environmental] transformations are "acceptable" ought to be relegated to a political process, be it a voting system, an inquiry commission, or whatever. The approach advocated here is thus directly opposed to those which incorporate value judgments about the desirable states of nature. . . . Rather than opt for simplistic index numbers for environmental quality, we prefer an information system relatively free of such value judgements. In fact, "environmental quality" is not a relevant concept except in the context of very carefully specified human preferences. (1980:27)

Although these comments refer generally to environmental monitoring, they underscore the difficulty of examining a "socionatural" problem like threats in a systematic and objective way. The sparsity of the literature and lack of comparative international data show the importance of our original question: What are the threats to national parks around the world? Also underscored is the need for a reliable method of gathering information about threats.

CHAPTER 3

Human Ecology and Parks

Just as quantitative data on threats to parks are unavailable, so, too, a clear and useful theoretical perspective has not emerged from the literature that deals with threats. The definition of threats has varied from study to study, as have the categories used to organize data.[1] Little or no attention has been paid to variables that might be correlated with the occurrence of threats or specific kinds of threats. Hypotheses have been neither proposed nor tested.

As the discussion in Chapter 2 suggests, threats result from interactions among a complex set of biological and social factors. In this chapter, a human ecological perspective is described and then applied to parks to facilitate a better understanding of these interactions and to suggest key variables in examining threats to parks.

THE HUMAN ECOLOGICAL PERSPECTIVE

In its most basic form, human ecology[2] is the study of the relationship between human beings and their environment (Theodorson and Theodorson 1969); the roots of this perspective lie primarily in general ecology, sociology, and anthropology (for a comprehensive literature review see Bruhn 1974; Micklin 1977). Frederick Clements's influential work *Plant succession* (1916) began the formal development of ecological principles, described in a series of general treatises on ecology (Dice 1952; Shelford 1963; Odum 1971; Krebs 1972). The ecologists' work was soon applied to human activity; sociologists helped spearhead the effort. Roderick McKenzie's article, The scope of human ecology (1926), Robert Park's Human ecology (1936), and Amos Hawley's *Human ecology: A theory of community structure* (1950) helped develop the central issues; other sociologists, such as Otis Dudley Duncan (1964), began to clarify central concepts.

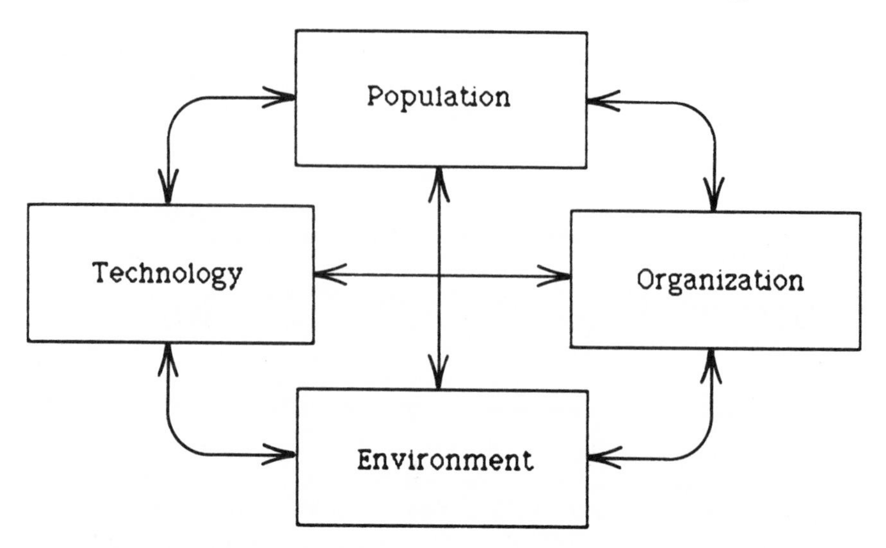

Figure 3.1. Key variables of the human ecosystem

Simultaneously, anthropology found ecology a useful theoretical and descriptive tool. Although Julian Steward coined the term *cultural ecology* in 1937, his later work (1955) led to an integration of the empirical field studies done by anthropologists. Recently, the anthropological version of human ecology has been refined in J. W. Bennett's *The ecological transition: Cultural anthropology and human adaptation* (1976) and questioned as well (Moran 1984b).

The essence of this general approach is a recognition of *Homo sapiens* as a part of Nature; the human ecosystem is its basic unit of analysis. The human ecosystem is defined by the interaction of population, social organization, and technology in response to a set of environmental conditions (Duncan 1964; Dunlap and Catton 1983; see Figure 3.1). These are in effect human ecology's "master variables"; their interaction is human ecology's major concern. Hawley notes: "Human ecology, which is also interested in the relations of man to his geographic environment, fastens its attention upon the human interdependencies that develop in the action and reaction of a population to its habitat" (1950:72).

Human ecosystems are not merely systems with people added as an afterthought; the uniqueness of *Homo sapiens* (the second assumption in Chapter 1) requires special analytical care. Bennett writes:

> If human activities are to be inserted into ecosystems, the system itself has to be reconceptualized: it is not a matter of a "natural" system being invaded by humans, but a complex whole system involving an interaction

> between the physical resources, animal species, and the human activities. This requires a shift in values as well: human components must be viewed as *analytically* equal to environmental components. (1984:302)

For the human ecologist, the behavior of aggregates is of special interest—whether it is the study of Tsembaga tribesmen (Rappaport 1967), the analysis of Chicago neighborhoods (Suttles 1972), or efforts to measure energy flow in industrial societies (Cook 1976). A. P. Vayda and R. A. Rappaport state that "the units important to ecologists are populations (groups of organisms living within a given area and belonging to the same species or variety), communities (all of the populations within a given area), and ecosystems (either individual organisms, populations, or communities, together with their non-living environments)" (1976:22).

Human ecosystems are dynamic and adaptive: The relations linking human populations to the environment can structurally change over time. As examples, new technology may lead to an alteration in settlement patterns (Cottrell 1951), and changed environmental conditions may cause a shift in the rhetorical uses of Nature (Burch 1971). As M. Micklin (1977) suggests, the analytic problem for ecological analysis is to explain these variations in adaptation. Human ecology asks: What conditions give rise to adaptive change? Why do some ecological units adapt more readily than others? What strategies of adaptation are available, and what are their consequences?

Adaptation is a crucial term here. The biologist's concept of adaptation has at least two meanings: (1) evolutionary genetic change, and (2) mechanisms used by organisms during their life spans to cope with the environment (Ricklefs 1973). Although evolutionary genetic change has undoubtedly been influential (see first assumption in Chapter 1), short-term adaptation within the human ecosystem is largely based on the importance of coping. Bennett writes:

> The rational or purposive manipulation of the social and natural environments constitutes the human approach to Nature: the characteristics of this style of adaptation must, it seems to me, become the heart of any approach to human ecology that concerns itself with the question of what people want and how they go about getting it, and what effects this has had on themselves and Nature. (1976:3)

A variety of examples—the Sioux Indian's adoption of the horse (Roe 1955), the cargo cults of the Pacific Islands (Jarvie 1963), and the recent energy crisis—reflect the response of human behavior to environmental and social change.

The human ecological perspective, like all others, has its limitations. Early efforts at human ecology overemphasized the importance of spatial relationships and often were blatantly deterministic in outlook (Young 1982; Ellen 1982). Biological metaphors such as niche, succession, energy flow, and species can be misused (Rosa and Machlis 1983). The human ecosystem is an attractive unit of analysis, but the definition of real-world system boundaries is often difficult to achieve. The simultaneous treatment of several variables makes analysis a complicated and rigorous activity; the amount of explanation, as compared to that of simple description, has been disappointingly small (Smith 1984).

Currently, the use of the ecosystem concept as an analytical tool has been increasingly questioned. E. F. Moran (1984b) suggests several problems: (a) a tendency to reify the ecosystem and give it properties of a biological organism; (b) an overemphasis on ecosystem self-regulation, even where it may not exist; (c) a static approach ignoring time and structural change; and (d) a lack of clear criteria for defining system boundaries. These problems can be reduced by minimizing the use of biological analogies, by allowing for unstable ecosystems to be examined, by analysis of ecological linkages over time, and by using system boundaries that have clear conceptual bases and can be delineated "on the ground."

Used as a descriptive tool, the human ecosystem remains robust because its flexibility allows the examination of both social and biophysical variables. Much of human behavior, from political conflict to energy conservation to forest management, can usefully be examined from this ecological perspective. This approach is practical because it is interdisciplinary and, as Bennett (1984) argues, it can help overcome the traditional academic divisions of the natural and social sciences. Moran, as part of his critique of the ecosystem concept, states: "Although it is no panacea for the difficulties of integrating the complex linkages between biotic and abiotic components of Nature, it is a helpful concept that reminds us that all systems are far more complex than our deductive or inductive models can ever conceive" (1984a:23).

THE HUMAN ECOLOGY OF PARKS

The human ecological perspective has been applied to urban parks (Dubos 1980), national parks (Machlis et al. 1981; Machlis and Field 1984), and wilderness areas (Darling 1969). The key components that make up a park, refuge, or preserve can be analyzed as subsystems of the park; the park is in turn embedded in a wider regional ecosystem and is influenced by the population, organization, technology, and environment that surround and interact with it (see Figure 3.2).[3] As

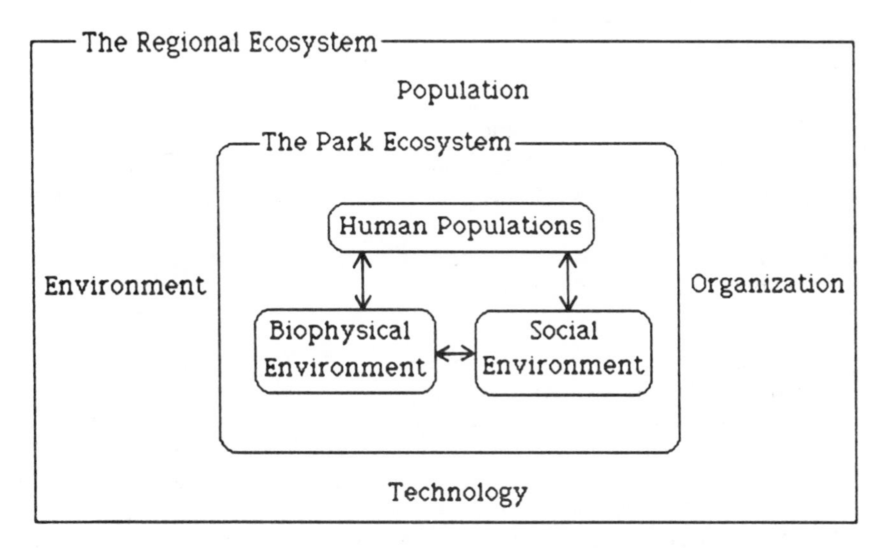

Figure 3.2. Conceptual diagram of park ecosystem

stated by W. J. Hart in *A systems approach to park planning,* "Within a given land area, all parks, no matter how large they may be, or for what purpose they were established, are related to each other, to the use of resources in the landscape which includes them, and to the society which supports them" (1966:XI).

The biophysical environment represents the natural resources of the park and the ecological processes necessary to sustain it. Included are the major energy flows (solar, fossil fuel, wind, and so on), the major ecosystem processes occurring in the park (erosion, succession, nutrient cycling, and so forth) and the plant and animal populations indigenous to the area. The characteristics of this environment may vary in different biomes. For example, the ecological processes required to sustain an ecosystem in an arid biome differ from those necessary in a tropical humid forest (Odum 1971). This variation in turn suggests that threats may differ in their intensity and effects from biome to biome.

The social organization of the park comprises several elements. Biosocial demands, such as the needs for sustenance, shelter, and education of the young, are a result of regularities in behavior imposed upon the species by evolutionary trends (Burch et al. 1978). These regularities can adversely affect park resources. For example, increasing demand for sustenance and shelter caused by population growth can accelerate local needs, and resources such as wood may be illegally removed from parks as adjacent lands become deforested.

Another component of the park social organization can be called human institutions—those formal patterns of social organization that enable human needs and wants to be satisfied (Martindale 1960). Institutions such as park management organizations may influence how threats are prevented, mitigated, and even perceived. The Man and the Biosphere Programme, for instance, emphasizes monitoring and research in protected areas such as parks (Risser and Cornelison 1979). Hence, parks included in this program may be more closely monitored for threats (USDI, NPS 1980).

An important element of the park ecosystem clearly is the human population. F. L. Campbell notes that

> humans are the dominant species in every National Park. As a result of our social evolution we have expanded into one niche after another. We have created new niches where none existed. Further, we are a highly generalized animal, capable of an immense range of behaviors. Our impact upon the complex biochemical systems that exist within the parks can be drastically altered as people take on new behaviors. In short, to understand the natural systems of the park you must understand the park's most dominant species. (1979:53)

In most parks, particularly in industrialized countries, this dominant population is park visitors. However, local inhabitants may be the primary population for parks in rural areas of less-developed nations that receive few visitors. The dominance of human populations in many parks suggests that understanding their resource demands and, therefore, their stage of economic development may have key importance in learning more about threats. This park ecosystem is in turn embedded in a wider regional ecosystem and is influenced by the population, organization, technology, and environment that make up that region.

The regional population's settlement pattern, migration rates, and age structure are likely to influence park use, as will the workings of regional organizations such as local and provincial governments, economic development commissions, large corporations, and nearby communities. Regional ecosystems are linked to park ecosystems—through such common features as airsheds, watersheds, electric power grids, and wildlife ranges. Regional technologies, particularly industrial plants and agriculture and transportation systems, may have significant impacts. Even national economic policies may influence park ecosystems. For example, an increase in international tourism to a country because of a favorable monetary exchange rate could increase park visitation and cause soil compaction in popular but fragile habitats. Hence, threats may originate both in the park and in the regional ecosystem.

If the components of a park system are further specified, several important subsystems emerge, including air, water, soil, vegetation, animal

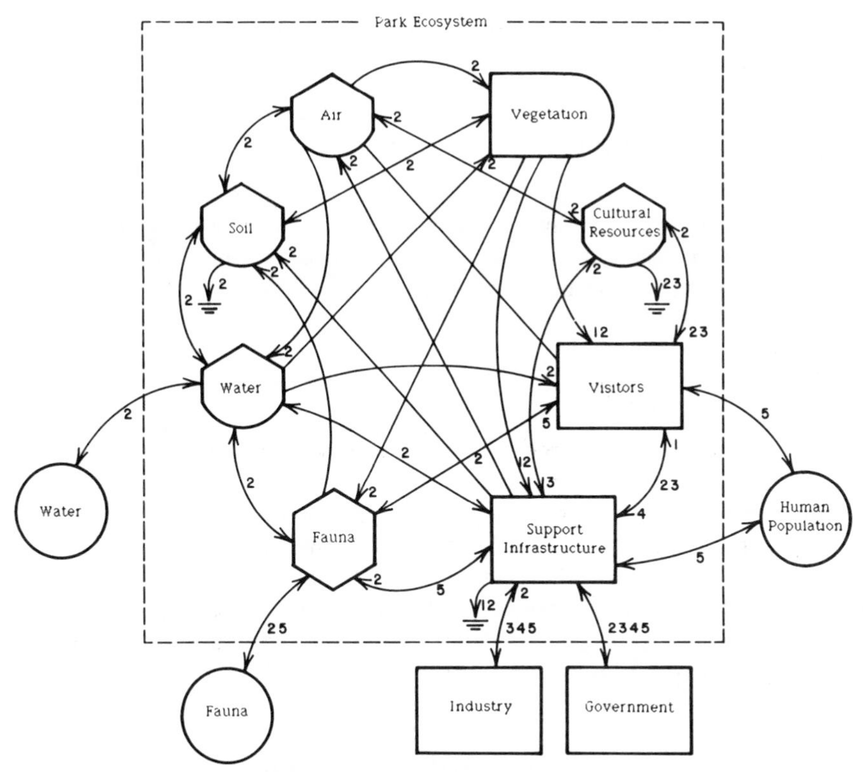

Figure 3.3. Specified diagram of park ecosystem. The notation is based on H. T. Odum's *Systems ecology: An introduction* (New York: Wiley and Sons, 1983). The shapes imply storage, production, and consumption functions, and the numbered arrows refer to flows of (1) energy, (2) materials, (3) information, (4) money, and (5) individuals. The ⏚ symbol represents a sink. Adapted from R. G. Wright and G. E. Machlis, Models for park management: A prospectus, Cooperative Park Studies Unit Report CPSU/UI SB85-1 (Moscow: University of Idaho, 1984), p. 20.

life, cultural resources, management and administration, and visitors (Tichnell et al. 1983). Figure 3.3 presents a conceptual model of such a park system and suggests key linkages via flows of energy, nutrients, money, and information (for a detailed discussion of the model, see Wright and Machlis 1984). Because complex linkages exist among these subsystems, threats associated with one subsystem may not impact all others similarly. Therefore, it is more useful to consider threats associated with particular subsystems, rather than with a park system as a whole. For example, organic pollution may cause unacceptable changes in the water subsystem, acceptable change in the soil subsystem, and little or no change in the cultural resources subsystem. Consequently, the threats facing a park are essentially the cumulative stresses upon each of the park's subsystems.

KEY STUDY VARIABLES

Several general variables (such as technology, environment, and social organization) have been suggested as important to the study of threats to parks. In this section we describe more precisely four variables drawn from this theoretical framework that form the core of our analysis: biome type, park size, stage of economic development, and management institutions.

Biome Type

As mentioned, the degree of impact to park ecosystems from a given threat may be related to the type of habitat found in them. For example, deforestation in tropical ecosystems often has more serious environmental consequences than it does in temperate zones (Hart 1966; Mueller-Dombois et al. 1984). R. F. Dasmann (1973; 1974), and later M.D.F. Udvardy (1975; 1984), designed a biogeographic classification scheme to help evaluate the ecosystem coverage of the world's protected areas. Terrestrial areas worldwide are assigned to one of the eight major realms, and 14 more specific biomes, as well as to the even more specific biogeographical provinces.

A key category is the biome, defined by Udvardy as

> groups of ecosystems that are related, and which show similarity in both appearance and internal structure, for they are influenced by the same climate, soil conditions, and elevational conditions, etc. Biomes are characterized by their dominant plant and animal members, but since animals are elusive, biome classification is based on the vegetational component of their stands. (1984:34)

Udvardy listed the following 14 principal biome types of the world:

1. Tropical humid forests
2. Subtropical and temperate rain forests or woodlands
3. Temperate needle-leaf forests or woodlands
4. Tropical dry or deciduous forests or woodlands
5. Temperate broad-leaf forests or woodlands, and subpolar deciduous thickets
6. Evergreen schlerophyllous forests, scrubs, or woodlands
7. Warm deserts and semideserts
8. Cold-winter (continental) deserts and semideserts

9. Tundra communities and barren arctic desert
10. Tropical grasslands and savannas
11. Temperate grasslands
12. Mixed mountain and highland systems
13. Mixed island systems
14. Lake systems

This classification exists only as a general guide because it is based on zonal ecosystems; specific habitats (such as lakes, rivers, and caves) are not considered. Thus parks may not encompass habitat types representing the biome in which they are found. Nevertheless, this system does provide the means for grouping together parks that have similar, general biotic characteristics. Biome type can be used as a general indicator of the biophysical environment found in a region. National parks located in a variety of different biome types were included in this study to examine the biophysical environment's influence on the characteristics of threats.

Park Size

The size and boundaries of national parks have been a management concern since the first parks were established. Early interest stemmed from human threats to park wildlife; for example, elk migrating outside Yellowstone National Park, United States, were shot by hunters (Albright 1918). Many national parks have been designated with inadequate ecological boundaries. Increasing human encroachment on parkland has made the ecological problems associated with these boundaries more apparent (Curry-Lindahl 1974b; Garratt 1984).

Nevertheless, guidelines concerning the size of parks did not exist until 1967. In that year the IUCN compiled the first list of national parks and equivalent reserves around the world, and in so doing introduced minimum-size criteria that parks would have to meet to be included. The numerical criteria were arbitrary and based on the population density of the country where a park was located: Parks in countries with more than 50 inhabitants per sq km had to be at least 500 ha; those in countries with fewer than 50 inhabitants per sq km were to be at least 2,000 ha (Forster 1973); islands, the exceptions to the rule, might be smaller. These size criteria were amended in 1969 when a minimum size of 1,000 ha was adopted, with the exception of ecological islands. This minimum-size criterion has been retained in subsequent editions of the list (IUCN 1982b).

In recent years, the increasing fragmentation of natural areas from surrounding lands has stimulated concern about the extinction of fauna within them (Soulé et al. 1979; Wilcox 1980; Western and Ssemakula

1981). Principles used to predict the relationship between species and area (that is, size) in ecological islands (MacArthur and Wilson 1967; Diamond 1975) have been applied to natural reserves such as parks to predict their ability to sustain wildlife populations or species diversity (Simberloff and Abele 1982; Bekele 1980; Willson and Carothers 1979; Miller 1978).

Although island biogeography may have specific applications in certain cases (such as Isle Royale in the United States or the Galapagos Archipelago in Ecuador), the island analogy has been extended to extremes. B. A. Wilcox writes that "virtually all natural habitats or reserves are destined to resemble islands, in that they will eventually become small isolated fragments of formerly much larger continuous natural habitat" (1980:95).

There are limits to the analogy. L. D. Harris (1984) notes that true islands are surrounded by a medium uninhabitable to terrestrial species and are often considered in the theory to be near a land mass that can provide a supply of new immigrants. Not all parks have these characteristics; in Harris's words "there is no 'continental' source pool of grizzlies to recolonize Mount Rainier National Park." Harris, using data from the northwest United States, has refined the island analogy, arguing for an "archipelago approach" that treats habitat islands as part of a wider system. In fact, scientific debate is being waged on the fundamental assumptions and practical applications of island biogeography (F. S. Gilbert 1980; Abele and Connor 1979). Current research, such as the Minimum Critical Size Project being conducted by the WWF in Brazil, represents the first large-scale field experiments designed to test and refine the application of species-area models.

Although authors generally agree that species are better off in larger reserves (Schonewald-Cox 1983; Diamond 1975), several studies conclude that size is not the only factor that should be considered. W. J. Boecklen and N. J. Gotelli (1984) examined several species-area models and found the application of these models to the design of nature reserves as "unwarranted." Precisely because parks are not islands and are permeable systems linked to regional environments by flows of energy, materials, populations, and even genetic information, the island metaphor is being increasingly scrutinized. Size and other design factors may in fact be subordinate to socioeconomic constraints in the surrounding region. D. Western and J. Ssemakula state: "Whether the savannah reserves become faunal islands will be decided by political and economic policies and practices rather than the principles of ecological design" (1981:7). Yet the size of a park or reserve remains a potentially useful variable in examining threats to parks, and hence parks of various sizes were included in our study.

Stage of Economic Development

Many authors (Blower 1984; Nelson 1978; Forster 1973; Hart 1966) have suggested that the type and intensity of threats may be influenced by the socioeconomic characteristics of the region where the park is located. Economic development may powerfully impact conservation activities. M. Clawson states:

> Countries vary greatly in their stage of economic development, as everyone knows. But this affects national parks in several ways. Higher average national incomes mean more people can afford to visit national parks, hence visitation rates will be high and rising. Such countries can better afford, if they try, to administer national parks well. In low income countries, there may well be a conflict between reservation of land for national parks and economic development. (1973:29)

But what is meant by economic development? Like threats, economic development is an amorphous concept capable of absorbing and reflecting the cultural values of those who use the term. The traditional definition of economic development has centered upon the sustained growth in a nation's gross national product (GNP). G. M. Meier and R. E. Baldwin state: "Economic development is a process whereby an economy's real national income increases over a long period of time" (1961:2). Also included in such traditional definitions has been the planned transfer of labor from the agricultural to manufacturing and service sectors (Eicher and Staatz 1983; Bhagwati 1966; Clark 1951).

From this narrow definition of development came in the 1950s and early 1960s several theories about stages of economic growth, represented by that of W. W. Rostow (1971). Rostow suggested that as a rule societies make a transition through general stages of economic growth: "It is possible to identify *all* societies, in their economic dimensions, as lying within one of five categories: the traditional society, the pre-conditions for take-off into self-sustaining growth, the drive to maturity, and the age of high mass consumption" (1971:1).

What is needed for takeoff is just the right combination of savings, investments, and aid; underdeveloped nations can follow the economic trajectory of already industrialized nations. These assumptions were made more explicit by such post-Keynesian analysis as the Harrod-Domar growth models (Meier and Baldwin 1961).

The theories about stages of economic growth were increasingly criticized in the late 1960s and early 1970s for several reasons: Savings and investments were necessary but not sufficient tools for development; the theories neglected the role of colonial exploitation in Western development, they failed to deal with the real conditions facing Third

World nations, and the accepted definition of development changed (Baran and Hobsbaum 1961; Kitching 1982).

In place of narrow definitions of economic development based upon national gains in GNP, these newer conceptions treat development as "a multi-dimensional process involving the reorganization and reorientation of entire economic and social systems" (Todaro 1981:56). With this new definition, structuralist models of development have increasingly been championed, which treat underdevelopment in terms of international and domestic power relationships and emphasize institutional constraints to economic growth. In these models, Third World development cannot proceed like that of the industrialized nations, precisely because of the structural dominance of industrialized nations over countries of the Third World.

In both the stages-of-development and structuralist arguments, the inverse relationship between the percentage of the labor force in agriculture and economic development is accepted, though with varying emphasis. To both, the percentage of a country's labor force in agriculture is one indicator of its economic development.[4]

Although these differing perspectives provide competing explanations for both the causes and remedies of underdevelopment, neither deals directly with the ecological consequences of development itself. Such consequences are often significant, as R. F. Dasmann et al. (1973) and more recently R. A. Carpenter (1983) point out. For example, J. E. Maragos et al. (1983) link development activities in tropical coastal ecosystems with potential change in species composition, lowered species diversity, reduction of standing stock, reproduction/recruitment failure, mass kills, inhibition of photosynthesis, and increased food chain concentration. Clearly, economic development is a variable that is likely to influence threats to parks.

Management Institutions

Two current major international programs directly involve the management of national parks. The first, created by the World Heritage Convention, allows member countries to submit natural or cultural areas of outstanding universal value for potential designation as world heritage sites (for an overview of the program, see Hales 1984). Members pledge to "identify, protect, and transmit to future generations" these areas, and grant rights of observation to other signatories. Technical assistance and related financial support are to be provided to members (Wetterburg 1984).

Adopted in 1972, the convention is a legal treaty among nations who are members of the UN Educational, Scientific, and Cultural Organization (UNESCO). By 1985, 83 nations have ratified the convention, and over

190 natural and cultural areas have been brought into the program. Many of the natural areas are previously established national parks.

The other program, titled Man and the Biosphere (MAB), is a series of projects launched by UNESCO in 1970 to facilitate rational land use and management. Project 8—The Conservation of Natural Areas and the Genetic Material They Contain—calls for the establishment of areas to conserve representative examples of the world's ecosystems and genetic diversity. Areas included in the project are referred to as biosphere reserves.

B. von Droste zu Hülshoff describes the goals of these areas:

> A biosphere reserve constitutes a representative example of one of the world's major ecosystems and thus conserves genetically viable plant and animal populations in their natural habitats; provides sites for long-term research on the structure, functioning and dynamics of ecosystems; combines research and monitoring, environmental education, training and demonstration; and seeks the support and participation of the local people, in part through research contributing to their social and economic development. A biosphere reserve is not an enclosed, inaccessible sanctuary. It blends with the surrounding landscape. (1984:690)

Currently 243 biosphere reserves have been designated in more than 60 countries, and in the large majority the special management program is superimposed upon preexisting national parks.

Parks designated as biosphere reserves or world heritage sites may illustrate the effects of certain specialized management programs on the status and extent of threats. Caution is needed because discrepancies may exist between what these conventions propose and what they actually accomplish. For example, although one of the responsibilities of the World Heritage Convention is to aid sites financially, no mechanism currently exists to determine which sites are in the greatest need (Hales 1984). Nevertheless, greater awareness and monitoring in these areas seem likely, and threats to these important areas may need special attention. In the next chapter, the methods used to gather data on these variables, along with the results of the survey, are described.

NOTES

1. For example, the NPCA study (1979) provides data on several "resources" adversely affected—water, air, silence, wildlife, and general biotic conditions. The categories are neither mutually exclusive nor sufficient, and the inclusion of silence is problematic. The IBP study is even more atheoretical in that respondents simply described problems and these descriptions were generalized.

2. Portions of this section are derived from G. E. Machlis, D. R. Field, and F. L. Campbell, The human ecology of parks, *Leisure Sciences* 4(3):195–212.

3. As described, the typical problems arise in systems analysis: Boundaries are sometimes arbitrary and difficult to define, linkages may be complex, and the structure of the system may change over time. For a general discussion of these problems, see J. G. Miller (1978).

4. There are several limitations to this economic indicator. Agricultural production is certainly not limited to food, and industrialization may include agricultural practices, such as plantation forestry (Kitching 1982). Nevertheless, the indicator is a broadly accepted one.

CHAPTER 4

The International Survey

The results of our study are based on the responses to an international survey of national park managers. Two objectives guided the research: (1) to identify the sources, extent, and effects of threats to the resources of a selected international sample of national parks; and (2) to examine whether the identified threats vary among parks of different sizes, in different biomes, in countries at different stages of development, and affiliated or not with special management programs. The research methods are detailed in Appendix A, and a brief, nontechnical description is provided here. The overview should aid the reader in evaluating the usefulness of the survey results.

THE RESEARCH METHODS EXPLAINED

The first step was to define more carefully the key variables of biome type, size, stage of economic development, and management institutions. Parks were categorized into one of four size categories, ranging from less than 10,000 ha to more than 1,000,000 ha, following the scheme of C. M. Schonewald-Cox (1983) and were assigned to one of M.D.F. Udvardy's (1975) 14 biome types (discussed in Chapter 3).

World Bank data were used to classify national parks. Those in countries with 20 percent or less of their labor force in the agricultural sector were categorized as "more developed," those in countries with more than 20 percent and less than 80 percent as "developing," and those with 80 percent or more as "less developed" (World Bank 1981). Finally, national parks involved with the Man and Biosphere and/or World Heritage programs were considered as affiliated with special management programs.

Only a sample of the world's national parks was surveyed. The choice of that sample was critical because an unrepresentative group of parks would lessen the usefulness of the results. To achieve a relatively

Table 4.1
Percent Biome Distribution of Sample Versus All Parks

Biome Type	Percent of All Parks (n=1,028)[a]	Percent of Sample (n=135)	Percent Difference
1. Tropical humid forests	14	10	-4
2. Subtropical and temperate rain forests or woodlands	10	9	-1
3. Temperate needle-leaf forests or woodlands	6	4	-2
4. Tropical dry or deciduous forests or woodlands	19	27	+8
5. Temperate broad-leaf forests, scrubs or woodlands, and subpolar deciduous thickets	4	5	+1
6. Evergreen sclerophyllous forests, scrubs or woodlands	13	7	-6
7. Warm deserts and semideserts	9	10	+1
8. Cold-winter (continental) deserts and semideserts	2	1	-1
9. Tundra communities and barren arctic desert	1	1	0
10. Tropical grasslands and savannas	2	2	0
11. Temperate grasslands	2	4	+2
12. Mixed mountain and highland systems	15	17	+2
13. Mixed island systems	3	3	0
14. Lake systems	1	0	-1

[a]The actual number of parks is 1,029, but data were missing for one park. Percentages may not equal 100 percent because of rounding.

Source: Compiled from International Union for the Conservation of Nature, List of biosphere reserves, IUCN Bulletin 13, 7-8-9(1982):69

homogeneous sample with regard to management objectives, only areas that met national park criteria as defined by the IUCN General Assembly in 1969 (see Chapter 1) were surveyed. Other categories of protected areas were not included.

A major objective was to select a sample of parks whose biome distribution would represent that of all national parks. Deriving a sample in this manner can help eliminate the potential bias resulting from oversampling a particular biome, as occurred in the IBP study (Clapham 1980). The biome distribution of the chosen sample is illustrated in Table 4.1; parks in tropical dry or deciduous forests are somewhat underrepresented. Similar efforts were made to have the sample vary in size, stage of economic development, and management affiliation. The final sample consisted of the 135 parks listed in Table 4.2.

A questionnaire was developed to survey managers at each of the selected parks. The questionnaire was designed around the theoretical framework described in Chapter 3. Park resources were grouped into seven main subsystems: water, air, soil, vegetation, animal life, management and administration, and a miscellaneous "other" category (subsystems are formally defined in Appendix A), and managers were asked a series of questions about a list of threats to each subsystem. After the questionnaire was tested at the 1982 World National Parks

Table 4.2
Parks Included in Survey Sample by Country

Australia
1. Croajingolong
2. Flinders Ranges
3. Hamersley Range
4. Kakadu
5. Kosciusko
6. Lower Glenelg
7. South West

Bahamas
8. Exuma Cays

Benin
9. "W"

Bolivia
10. Ulla Ulla National Faunal Reserve

Botswana
11. Chobe
12. Gemsbok

Brazil
13. Aparados da Serra
14. Brasilia
15. Iguacu
16. Serra de Bocaina

Cameroon
17. Waza

Canada
18. Banff
19. Glacier
20. Jasper
21. Kejimkujik
22. Nahanni
23. Pacific Rim
24. Riding Mountain
25. Waterton Lakes

Central African Republic
26. Bamingui-Bangoran
27. Manovo-Gounda-Saint Floris

Chile
28. Fray Jorge
29. Los Paraguas and Conguillio
30. Puyehue

Colombia
31. Cueva de los Guacharos
32. Puracé
33. Tayrona

Congo
34. Odzala

Costa Rica
35. Santa Rosa

Czechoslovakia
36. Krkonose

Denmark
37. Greenland

Ecuador
38. Galapagos

Ethiopia
39. Awash
40. Simien Mountains

France
41. Cévennes
42. Pyrénées occidentales
43. Vanoise

Gabon
44. Wonga-Wongué

Ghana
45. Mole

Guyana
46. Kaieteur

Hungary
47. Hortobagy

India
48. Bandipur
49. Kaziranga

Indonesia
50. Udjung Kulon

Italy
51. Abruzzo
52. Gran Paradiso

Ivory Coast
53. Marahoué
54. Mont Peko
55. Taï

Jamaica
56. Montego Bay
57. Ocho Rios

Japan
58. Daisetsuzan
59. Nikko

Kenya
60. Amboseli
61. Meru
62. Mount Kenya
63. Nairobi
64. Tsavo

Malawi
65. Kasungu
66. Lengwe
67. Nyika

Malaysia
68. Bako
69. Kinabalu

Mauritania
70. Banc d'Arguin

Nepal
71. Royal Chitwan
72. Langtang
73. Rara
74. Sagarmatha

Netherlands Antilles
75. Washington

New Zealand
76. Fiordland
77. Mount Cook
78. Nelson Lakes
79. Urewera

Niger
80. "W"

Norway
81. Rondane

Pakistan
82. Kirthar
83. Lal suhanra

Papua New Guinea
84. McAdam

Paraguay
85. Ybyku'i
86. Cerro Cora

Peru
87. Manu

Portugal
88. Peneda-Gerês

Rwanda
89. Kagera
90. Volcanoes

Senegal
91. Delta du Saloum
92. Niokolo-Koba
93. Djoudj

South Africa
94. Bontebok
95. Golden Gate Highlands
96. Kalahari Gemsbok
97. Kruger
98. Tsitsikama Forest & Coastal

Sri Lanka
99. Gal Oya
100. Wilpattu
101. Yala/Ruhuna

Sweden
102. Padjelanta
103. Peljekaise

Switzerland
104. Swiss

Tanzania
105. Serengeti

Thailand
106. Khao Yai

Tunisia
107. Ichkeul

Turkey
108. Gelibolu Peninsula
109. Köprülü Canyon

Uganda
110. Ruwenzori

USSR
111. Gauya State Reserve

United States
112. Big Bend
113. Bryce Canyon
114. Cape Cod
115. Death Valley National Monument
116. Everglades
117. Grand Canyon
118. Great Smoky Mountains
119. Mesa Verde
120. Olympic
121. Padre Island
122. Yellowstone
123. Yosemite

Upper Volta
124. "W"

Uruguay
125. Cabo Polonio
126. Santa Teresa National Monument

Venezuela
127. Canaima
128. Sierra Nevada
129. Yurubi

Zaire
130. Virunga

Zambia
131. Kafue
132. Mosi-Oa-Tunya
133. Kasanka

Zimbabwe
134. Ngezi
135. Wankie

Congress in Bali, Indonesia (see Tichnell et al. 1983), and prepared in English, Spanish, and French, copies were mailed to the selected parks in April 1983.

Limitations

Like all surveys, our effort to gather data on threats to parks has limitations that should be acknowledged. The most significant limitation is that the survey measured managers' perceptions of threats rather than the actual threats. These perceptions may not reflect the actual kind, extent, or seriousness of problems that face a specific park and may be affected by such variables as a respondent's education and values or by the level of monitoring in a park. For example, A. R. Clapham (1980:184) found a significant time dimension to responses in the IBP study: "It is perhaps not surprising that countries tend to emphasize current and quite recent impacts and seem to have overlooked those of the more distant past. . . ." Yet as a measure of what managers perceive to be threats, the survey is reasonably valid. Terms like *threats,* which could cause confusion, were carefully defined. Extensive testing helped to identify possible problems with definitions in the questionnaire wording, as well as other problems in the layout and design.

A second limitation is that this study was cross-sectional, conducted at one point in time (1983). Park conditions change and managers are transferred. If questionnaires were distributed at another time, and completed by a different set of managers, the results would undoubtedly vary. G. E. Machlis and R. G. Wright (1984) found in testing a similar questionnaire for U.S. national parks that the variation in responses among managers in the same park was high, even among those with similar backgrounds. Such variation may bias the results. For this reason, questions concerning the characteristics of respondents were analyzed to provide a profile of the respondents.

A third limitation is that threats to cultural resources such as buildings, archaeological sites, and trails are excluded from the study because of time, funding, and expertise restrictions. Cultural resources play an important role in many parks and may influence what threats occur to natural resources. Their exclusion limits the scope of our study and may result in certain threats going unreported.

Response to the Survey

Personnel in 100 of the 135 parks in the sample returned questionnaires, representing 49 countries. The response rate was 78 percent.[1] Table 4.3 shows that parks in more-developed countries had the highest response rate, 82 percent, compared to 66 percent for developing countries and 65 percent for less-developed nations. Response rate was similar for

Table 4.3
Response Rate by Stage of Economic Development and by Affiliation with Special Management Programs
(N=number of respondents, 98)

	Respondents	
	N	%
Stage of economic development		
Less developed	15	65
Developing	37	66
More developed	46	82
Special management program		
Nonaffiliated	73	72
Affiliated	25	76

parks with special management programs (76 percent) and unaffiliated parks (72 percent).

Table 4.4 shows the responses by biome type. Response was highest for parks in subtropical rain forests, cold winter continental deserts, evergreen schlerophyllous forests, and mixed mountain biomes. Response was lowest from tundra, tropical grasslands, and mixed island systems. More important, response was high for those biomes that include the largest number of national parks. The five most represented biomes (tropical humid forests, temperate rain forests, temperate forests, evergreen schlerophyllous forests, and mixed mountain systems) account for 71 percent of the world's national parks, and, hence, high response rates for these biome types are needed if results are to be generalized. As the table shows, response rates for these critical biomes were especially high (between 83 and 100 percent), except for the moderate return (54 percent) from parks in tropical humid forests.

A Profile of the Respondents

A majority of the respondents (72 percent) were involved in day-to-day park operations. Forty-eight percent indicated by their job titles (for example, park warden and park superintendent) that they were in charge of managing the area from which they responded. Another 24 percent were employed at the park in such capacities as ecologists or resource management technicians. Of the remaining respondents, 22 percent indicated they were employed at the regional or central government level in a capacity such as park director; 6 percent failed to indicate their job titles.

The managers generally had high levels of experience, education, and training. Indeed, 50 percent of the personnel involved in daily park operations indicated they had worked at least four years at the park

Table 4.4
Response Rate by Biome Type (N=number of respondents, 98)

Biome Type	Respondents N	%
1. Tropical humid forests[a]	7	54
2. Subtropical and temperate rain forests or woodlands[a]	12	100
3. Temperate needle-leaf forests and woodlands[a]	5	83
4. Tropical dry or deciduous forests or woodlands	23	64
5. Temperate broad-leaf forests or woodlands, and subpolar deciduous thickets	5	71
6. Evergreen schlerophyllous forests, scrubs or woodlands[a]	8	89
7. Warm deserts and semideserts	9	64
8. Cold winter continental deserts and semideserts	2	100
9. Tundra communities and barren arctic desert	0	0
10. Tropical grasslands and savannas	1	33
11. Temperate grasslands	4	80
12. Mixed mountain and highland systems[a]	20	87
13. Mixed island systems	2	50

[a]Major portion of world's national parks are included in these biomes.

whereas a comparative few (13 percent) indicated they had park experience of one year or less. A majority (54 percent) indicated they had 10 years or more of total experience in park management or related fields. With two exceptions, all had secondary educations, and 76 percent had completed some university work. Eighty-three percent of the managers said they had specialized training in park management or related fields.

Hence, a large majority of the parks responded; the response was well distributed along key study variables; and most respondents were involved in day-to-day park operations, with high levels of experience, education, and training.

GENERAL RESULTS

The Diversity of Threats to Parks

The survey data suggest that substantial and diverse threats confront the natural resources of national parks worldwide. On the questionnaire, 31 percent of all responses identified a total of 1,611 threats as being reported (identified as suspected or documented). The remaining valid responses identified specific threats as absent (50 percent) or unknown (3 percent).[2] Seventy-seven additional threats were written in by managers.

Table 4.5 presents the general results, showing the number of respondents reporting each threat listed on the survey. The results are listed by subsystem, and the threats are ranked within each subsystem by the frequency with which they were reported. Some threats clearly were more common than others; fully 76 percent of the parks reported illegal removal of animal life. The occurrences of some threats were relatively rare: Only 4 percent of the parks reported odors and 6 percent water temperature changes. Many of the threats were reported by one-quarter to one-half of the parks.

Table 4.6 lists the threats written in by respondents, again organized by subsystem. The list is revealing because certain threats—deforested watersheds, disease, volcanic activity—seem overwhelming. Others reflect the tolerance ranges of ecosystems; fire and lack of fire, too much rain and lack of flooding—all are on the list. Still other threats are site-specific and tractable: dumping of coffee pulp, road construction, short off-season, and staff attitudes.

Many of the written-in threats illustrate that even small disruptions of a park system can sometimes have large ecological consequences and that such disruptions are often difficult to control. A research officer and acting warden from Africa considered the lack of diesel fuel as a threat:

> Most of the park's water is pumped from bore holes into pans. The engines and pumps are antiquated and subject to numerous breakdowns. We have no good, reliable vehicles to service the engines, and parts for vehicles and engines are difficult to come by. Diesel availability is subject to the country being able to [import], and there have been shortages in the past due to [resistance movements] blowing up the oil pipeline. The water supply is dependent on the diesel, and without this there will be a substantial die off of animals.

Only three managers reported no threats to their parks. One explained:

> The reason for lack of threats to the park is that it is surrounded by State Forest and Crown Land, and ocean on the side. Four townships adjacent to park, closest three miles, have each less than 1,000 population. . . . Remoteness from population areas and good management by government bodies concerned results in no threats to the park for forseeable future.

A Core of Threats

The survey points out that many threats are common to parks worldwide (see Table 4.7 for a list of the 10 most reported threats). The most common threat is illegal removal of animal life, reported by 74 of the 98 parks. The following response is typical:

> Poaching is widespread, exacerbated by the shape of the park, population density on its border, its position on the [frontier] and the dense vegetation on rugged terrain. Primary poaching targets are elephant, antelope and gorilla. Poacher's dogs, plus feral dogs from the surrounding farmlands, are now found throughout the park hunting on their own. . . . The major worry in [ ] National Park today is commercialized poaching for elephant and rhino. These two species are threatened because of high demand of ivory and rhino horn in outside countries. The poaching is done right inside the park. Law enforcement officers are challenged at times because they carry single-shot firearms as opposed to machine guns used by these sophisticated poachers.

Other commonly reported threats included lack of personnel (73 percent), removal of vegetation (61 percent), and erosion (58 percent). No threats were included on the list for two subsystems—air and water.

The list of most reported threats suggests a core of related resource problems facing national parks. The relationships among conflicting demands, local attitudes, lack of personnel, and illegal removal of animal life have been documented through several case studies (see Marks 1984, for example), and the survey data suggest that a majority of national parks face this particular combination of threats.

Suspected Versus Documented Threats

Although the problems facing national parks are substantial, our knowledge of them is sometimes inconclusive. The data suggest that many threats are not well documented. Table 4.8 shows the status of reported threats by subsystem: Sixty percent were documented whereas 40 percent were only suspected. Only modest variation took place by subsystems, with the lowest percentage of documented threats related to soil (52 percent) and the highest to vegetation (69 percent). The importance of documentation through ecosystem monitoring was stressed by several managers. A U.S. park biologist made the following suggestion:

> Monitoring of natural resources should be a programmed, professional activity in parks: managers cannot be expected to prevent or minimize threats without some recurrent means of detecting and evaluating them.

> Research programs should be funded to develop reliable, cost effective means of continually monitoring for threats.

Threatened Subsystems

All the subsystems reported threats; however, the animal life, management, and vegetation subsystems were most often affected. Table 4.9 indicates the number of reported threats for each subsystem. Although the percentage for each subsystem is influenced by the number of threats listed on the questionnaire, the three aforementioned subsystems accounted for 69 percent of all reported threats. Air was the only subsystem not threatened in more than 50 percent of the parks.

Man and Nature as Causes of Threats

Table 4.10 shows that 65 percent of all reported threats were indicated to be caused by man; when these are combined with those caused by man and Nature, humans are found to influence more than 75 percent of the reported threats in each subsystem. As might be expected, the percentage of man-caused threats is highest for the management subsystem (83 percent) and miscellaneous category (89 percent). Natural threats are highest for the water system (19 percent).

These numbers suggest that humans are the principal cause of threats to parks, though Nature plays an important role as well. Twelve percent of the reported threats were attributed solely to Nature. One manager from North America wrote:

> [The park] is still one of those rare parks that is located in a non-developed area. There is no road access and infrastructures in the park are limited to a few ranger cabins and portage trails. Threats listed are on a small scale. As a matter of fact, most of the threats can be attributed to nature, e.g., forest fires leading to loss of habitat and removal of vegetation.

Natural and man-caused threats may have differential impacts upon park ecosystems. A chief park warden wrote:

> It has been our experience that natural threats or catastrophes (even prolonged droughts and disease epidemics) in national parks, particularly those of considerable surface area, rarely cause the extinction of species or have lasting detrimental effects in the ecosystem. An exception here is overpopulation by destructive herbivores such as elephant, hippo and buffalo.

The data clearly illustrate the interplay of man and Nature in producing threats to parks; one-fifth of the total number of reported threats was attributed to both man and Nature. A wildlife manager officer from Africa provided an example:

> I would like to make clear that the removal of vegetation by nature is not due to natural overstocking. It is due to poaching pressure which squeezes the animals to one corner of the park and thus damages the vegetation.

The combined cause was highest for the soil subsystem (38 percent) and lowest for management (6 percent).

The Location of Threats to Parks

Threats originated from a combination of sources located both inside and outside park boundaries (35 percent), as well as from sources found exclusively inside (34 percent) or outside (24 percent) (see Table 4.11). Surprisingly, 69 percent of the reported threats at least partially originated within park boundaries.

As shown in Table 4.11, the sources of reported threats to the air and water subsystems were most frequently cited to be outside parks; threats to the soil subsystem were reported as mostly internal. Underscoring the theoretical framework in Chapter 3, several of the 10 most reported threats in the study—illegal removal of animal life, removal of vegetation, local attitudes, and conflicting demands—can be attributed to interactions between parks and socioeconomic regions adjacent to them.

The comments by park managers reflect the permeability of park boundaries. One park administrator in Central America gave a specific example:

> Chemical contamination is caused by a nearby farm where agricultural crops are planted, such as sorghum and cotton. Chemicals are applied to control weeds and insects. The fungicides are sprayed by airplane, much of which are air-blown to the park and deposited into water sources, land and vegetation.

Not only are boundaries permeable, in many parks their lack of definition represents a threat in itself. Thirty percent of all respondents reported undefined boundaries as a threat to management (see Table 4.5). A park superintendent wrote simply:

> Park management is hampered by the fact that this park has not yet been gazetted and therefore the National Parks Act and Regulations do not apply.

Most threats are localized within park areas. Table 4.12 shows that almost one-half of all reported threats impacted less than one-fourth of the park area. Similar results are shown when responses are analyzed by subsystem. Threats to soil and vegetation subsystems are the most limited, with 68 percent of the threats to soil subsystems affecting less than one-fourth of park area. As could be expected, the threats to air and management subsystems affect larger portions of park area. The kind of threat may greatly influence its physical distribution within park boundaries. As an Australian ranger noted,

> Fire caused by lightning is a problem to vegetation and also animal life. As it can strike anywhere in the park, the whole park area is at risk at some time. In 1982, some 19 lightning strikes have burnt about one-fifth of the park.

RESULTS BY KEY STUDY VARIABLE

Park Size

As discussed earlier, much of the literature on minimum habitat size argues that larger areas may provide better protection for species and ecosystem processes (Schonewald-Cox 1983). Our data suggest that threats reported to parks of different sizes were similar in status, cause, location, and park area affected. Some variation was found when subsystems were compared.

Most of the reported threats to parks in each size category were documented and were caused by man. The percentage of documented threats is similar for the smallest and largest size categories (68 and 69 percent, respectively), and little variation can be seen in cause of threats between these categories. The cause of threats varies more among subsystems; for example, Table 4.13 shows that four subsystems (water, soil, vegetation, and animal life) in the largest parks have a greater percentage of naturally caused threats than they have in parks in any of the other size categories.

As Table 4.14 illustrates, larger parks had a slightly greater percentage of threats originating inside their boundaries than did smaller ones. Some patterns emerge at the subsystem level: The percentage of external threats to the animal life subsystem is highest for the smallest parks (38 percent) and decreases with size; in contrast, the percentage of internal threats to the soil subsystem increases with size. An anomaly is revealed: None of the 12 smallest parks reported threats to the air subsystem.

Table 4.5
Threat Ranks by Subsystem (N=number of reported threats, 1,534)[a]

Rank	Threat	Respondents Reporting N	Respondents Reporting %
Subsystem: Water			
1	Increased demand	31	32
2	Silt	25	26
2	Chemical pollution	25	26
3	Inadequate rain	22	22
3	Blocked river or stream flow	20	20
4	Temperature change	6	6
Subsystem: Air			
1	Chemical pollution	19	19
2	Dust	12	13
3	Smoke	11	11
Subsystem: Soil			
1	Erosion	57	58
2	Inadequate cover of vegetation	36	37
3	Compaction	26	27
4	Loss of soil nutrients	20	20
5	Chemical pollution	10	10
6	Increased concentration of salts	7	7
Subsystem: Vegetation			
1	Removal of vegetation	60	61
2	Fire	50	51
3	Trampling	46	47
4	Exotic plants	42	43
5	Exotic animals	36	37
6	Plant succession	32	33
7	Inadequate water supply	19	19
8	Flooding	18	18
9	Chemical pollution	11	11
Subsystem: Animal Life			
1	Illegal removal	74	76
2	Human harrassment	49	50
3	Loss of habitat	47	48
4	Fire	42	43
5	Habitat change	37	38
6	Exotic animals	36	37
7	Disease	32	33
8	Inadequate supply of food	26	27
9	Overpopulation of specific species	25	26
10	Inadequate water supply	22	22
11	Blocked migratory routes	18	18
11	Legal removal	18	18
12	Chemical pollution	14	14
13	Flooding	11	11
Subsystem: Management			
1	Not enough personnel	72	73
2	Local attitudes	52	53
2	Conflicting demands	52	53
3	Unlawful entry	39	40
4	Undefined boundaries	29	30
5	Too many visitors	23	23
6	Unsafe conditions	20	20
7	Too much building development	14	14
Subsystem: Other			
1	Degraded scenic viewpoints	37	38
2	Litter	36	37
3	Mining	31	32
4	Noise	19	20
5	Degraded geological features	14	14
6	Odors	4	4

[a]Seventy-seven threats written in by respondents are not included.

Table 4.6 Threats Written in by Respondents

Rank	Threat	Number of Respondents Reporting
Subsystem: Water		
1	Organic pollution	5
2	Reduced water flow	2
3	Hydroelectric dams	2
4	Exotic water plants	2
5	Disease	2
6	Unnatural water delivery (time of year)	1
7	Dumping of coffee pulp	1
8	Land development	1
9	pH decrease	1
10	Reduced hydrothermal discharge	1
11	Exotic water animals	1
12	Diesel availability (for water pumps)	1
13	Deforestated watershed	1
14	Too much rain	1
15	Drying lakes and meadows	1
16	Lack of silt in water	1
17	Lack of watering points	1
18	Evaporation	1
Subsystem: Soil		
1	Road construction	2
2	Unauthorized off-road vehicle use	1
3	Fire	1
4	Laterization	1
Subsystem: Vegetation		
1	Insects	3
2	Weather extremes	1
3	Disease	1
4	Unauthorized off-road vehicle use	1
5	Treated vegetation not responding	1
6	Actions of local authorities	1
7	Overpopulation of animals	1
8	Lack of fire	1
9	Lack of flooding	1
10	Irrigation	1
11	Red deer	1
Subsystem: Animal life		
1	Insufficient area	2
2	Excessive shellfish harvesting	1
3	Cropping	1
4	Anchor damage to reefs	1
5	Native species extinction	1
6	Short off-season	1
7	Lack of flooding	1
8	Roaming dogs	1
9	Intensive animal breeding	1
Subsystem: Administration and management		
1	Lack of funding	3
2	Lack of equipment	2
3	Lack of trained personnel	2
4	Lack of infrastructure	2
5	Not legally gazetted	1
6	Insufficient facility maintenance	1
7	Native claims	1
8	Political pressure	1
9	Unfavorable staff conditions	1
10	Staff attitudes	1
11	Lack of professional service at district level	1
12	Political-economic conditions	1
13	Lack of agency coordination	1
Subsystem: Other		
1	Volcanic activity	1
2	Archaeological resource degradation	1
3	Loss of historic resources	1
4	Coastal zone construction	1
5	Cultural resources degradation	1
	Total	77

Table 4.7
Ten Most Reported Threats (N=number of parks reporting, 98)

Subsystem	Threat	N	%
Animal life	Illegal removal of animal life	74	76
Management	Lack of personnel	72	73
Vegetation	Removal of vegetation	60	61
Soil	Erosion	57	58
Management	Local attitudes	52	53
Management	Conflicting demands	52	53
Vegetation	Fire	50	51
Animal life	Human harrassment	49	50
Animal life	Loss of habitat	47	48
Vegetation	Trampling	46	47

Table 4.8
Status of Reported Threats (N=number of threats, 1,611)

	Status			
	Suspected		Documented	
Subsystem	N	%	N	%
Total	645	40	966	60
Water	56	36	99	64
Air	19	45	23	55
Soil	78	48	83	52
Vegetation	100	31	227	69
Animal life	212	46	249	54
Management	111	35	208	65
Other	69	47	77	53

Table 4.9
Reported Threats by Subsystem
(N=number of threats, 1,611)

	Reported Threats	
Subsystem	N	%[a]
Total	1,611	100
Water	155	10
Air	42	3
Soil	161	10
Vegetation	327	20
Animal life	461	29
Management	319	20
Other	146	9

[a]Percentages do not equal 100% because of rounding.

Table 4.10
Cause of Reported Threats[a] (N=number of threats, 1,611)

	Cause					
	Man		Nature		Both	
Subsystem	N	%	N	%	N	%
Total	1,051	65	193	12	289	18
Water	88	57	30	19	30	19
Air	27	64	2	5	12	29
Soil	68	42	26	16	62	38
Vegetation	181	55	50	15	84	26
Animal life	292	63	73	16	75	16
Management	265	83	8	2	18	6
Other	130	89	4	3	8	5

[a]Row percentages may not equal 100% because of the exclusion of missing and invalid responses.

Table 4.11
Location of Reported Threats[a] (N=number of threats, 1,611)

	Location					
	Inside		Outside		Both	
Subsystem	N	%	N	%	N	%
Total	546	34	387	24	570	35
Water	33	21	62	40	52	34
Air	4	10	19	45	18	43
Soil	93	58	11	7	53	33
Vegetation	130	40	52	16	135	41
Animal life	141	31	101	22	187	41
Management	93	29	106	33	74	23
Other	52	36	36	25	51	35

[a]Row percentages may not equal 100% because of the exclusion of missing and invalid responses.

Table 4.12
Park Area Affected by Reported Threats[a] (N=number of threatened subsystems, 490)

	Area Affected							
	No more than 1/4		More than 1/4, Less than 1/2		More than 1/2, Less than 3/4		More than 3/4	
Subsystem	N	%	N	%	N	%	N	%
Total	240	49	80	16	47	10	67	14
Water	33	49	12	18	8	12	10	15
Air	11	41	0	0	1	4	9	33
Soil	44	68	8	12	5	8	5	8
Vegetation	48	56	21	24	8	9	6	7
Animal life	41	44	19	20	10	11	14	15
Management	36	41	14	16	13	15	18	21
Other	27	42	6	9	2	3	5	8

[a]Row percentages may not equal 100% because of the exclusion of missing and invalid responses.

A majority of the reported threats to parks in each size category except the 100,000–1,000,000 ha category affect less than one-fourth of a park (see Table 4.15). Again, the percentage of localized threats varies more at the subsystem level. For example, localized threats to the water subsystem decrease from the smallest parks (62 percent) to largest parks (43 percent). Yet the variation is not systematic; the percentage of localized threats to animal life is exactly the same for parks with less than 10,000 ha as for parks with more than 1 million ha.

Table 4.16 ranks the 10 most reported threats for parks in each size category. Several of the most numerous threats (for example, lack of personnel and illegal removal of animal life) are similarly ranked in each size category; others, such as litter, are listed in only one. Numerous threats in each size category were reported by the same number of respondents.

Biome Type

Not all the sampled biome types had large cell sizes. To make reliable comparisons among different biome types, the four biomes with the highest response rates were analyzed: subtropical and temperate rain forest or woodlands; tropical dry or deciduous forest or woodlands; warm deserts and semideserts; and mixed mountain and highland systems. These biomes contain 63 percent of the world's national parks.

Threats reported to parks in selected, different biomes were similar in status, cause, location, and park area affected. The majority of the threats reported by parks in each of these biomes were documented varying from 70 percent in subtropical forests to 50 percent in tropical forests (Table 4.17). More variation can be identified at the subsystem level; for example, every subsystem in the subtropical forest biome had a higher percentage of documented threats than the tropical forest type.

As Table 4.18 illustrates, most reported threats in every biome type were perceived to be caused by man; however, this perception varied from 78 percent for parks in subtropical forests to 54 percent for parks in warm deserts. Some differences are revealed: Parks in tropical forest and warm desert biomes reported more naturally caused threats than those in mixed mountain systems for every subsystem; the highest proportion of naturally caused threats (37 percent) was reported for the water subsystems of parks in tropical forest biomes.

Little variation occurs in the locations of reported threats among parks in the different biome types (see Table 4.19). Regardless of biome type, the water and air subsystems were mostly threatened by sources outside park boundaries. Some differences are identified among subsystems; for example, five of the subsystems (water, air, soil, animal life, and management) in the parks of the subtropical forest biome had a greater percentage of reported threats from sources outside park boundaries than did parks in the mixed mountain system. The soil subsystem had the greatest percentage of reported threats originating within parks for every biome type.

The 10 most reported threats for each biome type are listed in Table 4.20. Every biome type has at least 5 threats in common with another; however, the threats are often ranked differently. Removal of vegetation was commonly reported as a threat for parks in subtropical forests (75 percent) and mixed mountain systems (85 percent); it was less common in tropical forest (57 percent) and warm desert parks (56 percent). Several threats, such as mining, noise, and degraded scenic viewpoints, are listed for only one biome.

Stage of Economic Development

In the previous discussion, economic development and national park ecosystems were described as inextricably linked, especially in developing countries with few per capita surpluses and little arable land. The survey data suggest that threats reported to parks in countries at different stages of economic development varied in status, location, and cause, both generally and at the subsystem level. Parks in more-developed countries reported the highest percentage of documented threats (64 percent). And with the exception of management, parks in more-developed countries reported the highest percentage of documented threats for each subsystem as well.

A similar pattern is revealed for the cause of threats. As shown in Table 4.21, the percentage of threats linked to man increases from parks in less-developed countries (57 percent) to those in more-developed countries (70 percent), whereas threats resulting from Nature decrease slightly. At the subsystem level, the results are more varied. For example, the proportion of natural threats to water decreases from 42 percent in less-developed nations to 7 percent in more-developed countries. The air, soil, and animal life subsystems have a greater percentage of man-caused threats in both less-developed and more-developed countries than in developing nations.

Little variation occurs for other subsystems. The proportion of man-caused threats to animal life is 65 percent for parks in less-developed nations, 62 percent for those in developing nations, and 64 percent for those in more-developed nations. For the vegetation, animal life, management, and other categories, only slight differences in the percentage of man-caused threats can be seen between parks in developing and more-developed countries.

In comparison to less-developed nations, respondents from developing and more-developed countries reported approximately 10 percent more threats as having outside sources (see Table 4.22). Moreover, respondents in more-developed countries reported a greater proportion of reported threats from outside sources for every subsystem than in less-developed nations. Yet some of the differences are quite small and suggest that the location of threats to certain subsystems varies little by stage of economic development. For example, the percentages for internal threats to the animal life subsystem are 31 percent for less-developed, 28 percent for developing, and 32 percent for more-developed countries. Similar results are shown for the management subsystem.

A majority of the reported threats to parks in countries at each stage of economic development apparently affect less than one-half of the park area (see Table 4.23). Threats to certain subsystems show systematic variation: Water, soil, and animal life subsystems all show more localized threats in more-developed countries than in the less-developed nations. In developed countries, the most generalized threats are to the air subsystem; in the developing countries the most generalized threats are to the water subsystem; and in the less-developed nations, the most generalized threats are to management.

Several threats, such as lack of personnel and illegal removal of animal life, are common to parks in countries at all stages of development (see Table 4.24). The most numerous threats in less-developed and developing countries are the same, but most are ranked differently. There are some substantial variations: Although 87 percent of the parks in less-developed countries cited conflicting demands as a threat, only 41 percent of those in developing and 52 percent in developed countries reported such a threat. Several threats, such as degraded scenic viewpoints and trampling, common in more-developed countries, are absent from the other rankings.

The data suggest that the core of threats facing national parks will reflect regional and national development patterns and that as countries develop economically, threats may change. The implications for park management are discussed in Chapter 5.

Affiliation with Special Management Programs

The status, cause, and location of reported threats to affiliated and nonaffiliated parks were very similar. The percentage of documented threats is slightly lower for parks affiliated with special programs (57 percent) than nonaffiliated ones (61 percent); this pattern holds true for all subsystems except air. Variation in the cause of threats is less than 3 percent, with similar results for each subsystem.

A few differences are found in the location of threats, primarily at the subsystem level. Table 4.25 shows the location of threats for parks by their affiliation with special management programs. Interestingly, five subsystems (water, soil, vegetation, animal life, and management) in affiliated parks show more reported threats originating inside park boundaries than in nonaffiliated units. For example, the parks with special affiliation reported that 38 percent of the threats to management were inside the park; 25 percent of the nonaffiliated programs so reported.

Further variation is evident in Table 4.26, which identifies the park area affected by threats. In parks with special affiliations, threats were consistently reported as affecting a larger proportion of area. The pattern holds for each and every subsystem. For example, 32 percent of the threats to animal life in affiliated parks affect less than one-fourth of the park area, as opposed to 49 percent for nonaffiliated parks.

The 10 most reported threats are listed by affiliation in Table 4.27. The 3 most reported threats (illegal removal of animal life, lack of personnel, and removal of vegetation) are ranked identically in affiliated and nonaffiliated parks. Some less reported threats are ranked differently: Parks affiliated with special programs did not report local attitudes as one of their most common threats; for nonaffiliated parks this is the third most common threat, reported by 58 percent of the areas. The importance of local attitudes was stressed by an African chief park warden:

> There is a big need for wildlife education for adults; without this all efforts made to protect remaining animals will fail.

A park ranger from Oceania repeated the theme:

> Lack of publicity and education to villagers, schools, etc., about national parks, their values and benefits, have made people see that they don't benefit from national parks.

FURTHER EXPLORATION

The preceding discussion has shown that the status, cause, location, and area affected by reported threats differ for several of the key study variables. The analysis was primarily based on the percentage of national parks that reported threats, first to all subsystems and then to each subsystem. In this section we examine the specific threats in more detail. Rather than being an exhaustive or elaborate analysis, the material simply

Table 4.13
Cause of Reported Threats by Park Size[a] (N=number of threats, 1,611)

	Less than 10,000 ha (n=12)						10,000 through 99,999 ha (n=40)					
	Man		Nature		Both		Man		Nature		Both	
Subsystem	N	%	N	%	N	%	N	%	N	%	N	%
Total	126	65	19	10	34	18	407	65	72	12	121	19
Water	11	61	1	6	5	28	29	53	12	22	11	20
Air	0	0	0	0	0	0	13	68	1	5	5	26
Soil	7	35	3	15	10	50	26	38	12	18	30	44
Vegetation	21	51	7	17	9	22	81	59	20	14	33	24
Animal life	40	73	7	13	5	9	104	62	21	12	33	20
Management	32	76	1	2	3	7	110	86	3	2	8	6
Other	15	83	0	0	2	11	44	90	3	6	1	2

[a]Row percentages by park size may not equal 100% because of the exclusion of missing and invalid responses.

Table 4.14
Location of Reported Threats by Park Size[a] (N=number of threats, 1,611)

	Less than 10,000 ha (n=12)						10,000 through 99,999 ha (n=40)					
	Inside		Outside		Both		Inside		Outside		Both	
Subsystem	N	%	N	%	N	%	N	%	N	%	N	%
Total	66	34	65	34	51	26	183	29	150	24	244	39
Water	6	33	8	44	3	17	8	15	23	42	21	38
Air	0	0	0	0	0	0	2	11	10	53	7	37
Soil	8	40	4	20	7	35	34	50	3	4	31	46
Vegetation	16	39	9	22	16	39	45	33	21	15	68	49
Animal life	12	22	21	38	18	33	36	21	40	24	75	44
Management	15	36	18	43	4	10	42	33	40	31	24	19
Other	9	50	5	28	3	17	16	33	13	27	18	37

[a]Row percentages by park size may not equal 100% because of the exclusion of missing and invalid responses.

100,000 through 999,999 ha (n=38)						1,000,000 through 9,999,999 ha (n=8)					
Man		Nature		Both		Man		Nature		Both	
N	%	N	%	N	%	N	%	N	%	N	%
437	67	74	11	111	17	81	58	28	20	23	16
39	58	12	18	13	19	9	60	5	33	1	7
12	60	1	5	6	30	2	67	0	0	1	33
31	51	8	13	19	31	4	33	3	25	3	25
69	57	18	15	32	26	10	38	5	19	10	38
126	66	30	16	29	15	22	47	15	32	8	17
102	82	4	3	7	6	21	88	0	0	0	0
58	88	1	2	5	8	13	100	0	0	0	0

100,000 through 999,999 ha (n=38)						1,000,000 through 9,999,999 ha (n=8)					
Inside		Outside		Both		Inside		Outside		Both	
N	%	N	%	N	%	N	%	N	%	N	%
236	36	136	21	244	37	61	44	36	26	31	22
16	24	23	34	25	37	3	20	8	53	3	20
1	5	8	40	10	50	1	33	1	33	1	33
42	69	3	5	14	23	9	75	1	8	1	8
57	47	19	16	43	35	12	46	3	12	8	31
66	35	35	18	82	43	27	57	5	11	12	26
31	25	34	27	44	35	5	21	14	58	2	8
23	35	14	21	26	39	4	31	4	31	4	31

Table 4.15
Park Area Affected by Reported Threats by Park Size[a] (N=number of threatened subsystems, 490)

Subsystem	Less than 10,000 ha (n=12)								10,000 through 99,999 ha (n=40)							
	No more than 1/4		More than 1/4, Less than 1/2		More than 1/2, Less than 3/4		More than 3/4		No more than 1/4		More than 1/4, Less than 1/2		More than 1/2, Less than 3/4		More than 3/4	
	N	%	N	%	N	%	N	%	N	%	N	%	N	%	N	%
Total	37	61	10	16	4	7	4	7	97	51	34	18	22	11	21	11
Water	5	62	2	25	1	12	0	0	13	50	5	19	4	15	3	12
Air	0	0	0	0	0	0	0	0	8	53	0	0	1	7	2	13
Soil	8	89	1	11	0	0	0	0	16	64	2	8	4	16	2	8
Vegetation	9	75	3	25	0	0	0	0	14	42	12	36	4	12	2	6
Animal life	6	50	2	17	2	17	0	0	18	51	6	17	2	6	6	17
Management	4	40	2	20	0	0	4	40	14	40	8	23	6	17	4	11
Other	5	50	0	0	1	10	0	0	14	61	1	4	1	4	2	9

Subsystem	100,000 through 999,999 ha (n=38)								1,000,000 through 9,999,999 ha (n=8)							
	No more than 1/4		More than 1/4, Less than 1/2		More than 1/2, Less than 3/4		More than 3/4		No more than 1/4		More than 1/4, Less than 1/2		More than 1/2, Less than 3/4		More than 3/4	
	N	%	N	%	N	%	N	%	N	%	N	%	N	%	N	%
Total	80	41	34	18	17	9	37	19	26	60	2	5	4	9	5	12
Water	12	44	5	19	1	4	5	19	3	43	0	0	2	29	2	29
Air	2	18	0	0	0	0	7	64	1	100	0	0	0	0	0	0
Soil	15	58	5	19	1	4	3	12	5	100	0	0	0	0	0	0
Vegetation	19	58	5	15	4	12	3	9	6	75	1	12	0	0	1	12
Animal life	13	34	10	26	6	16	6	16	4	50	1	12	0	0	2	25
Management	13	38	4	12	5	15	10	29	5	62	0	0	2	25	0	0
Other	6	24	5	20	0	0	3	12	2	33	0	0	0	0	0	0

[a]Row percentages by park size may not equal 100% because of the exclusion of missing and invalid responses.

Table 4.16
Ten Most Reported Threats by Park Size[a] (N=number of parks reporting, 98)

Less than 10,000 ha (n=12)			10,000 through 99,999 ha (n=40)			100,000 through 999,999 ha (n=38)			1,000,000 through 9,999,999 ha (n=8)		
Threat	N	%	Threat	N	%	Threat	N	%	Threat	N	%
Illegal removal of animal life	10	83	Lack of personnel	28	70	Illegal removal of animal life	31	82	Fire (V)	7	88
Erosion	9	75	Illegal removal of animal life	26	65	Lack of personnel	28	74	Lack of personnel	7	88
Removal of vegetation	9	75	Removal of vegetation	24	60	Removal of vegetation	24	63	Illegal removal of animal life	7	88
Lack of personnel	9	75	Local attitudes	24	60	Erosion	22	58	Fire (A)	6	75
Fire (V)	8	67	Erosion	23	58	Loss of habitat	21	55	Inadequate rain (A)	5	62
Fire (A)	7	58	Conflicting demands	22	55	Exotic animals (A)	20	53	Human harrassment	5	62
Human harrassment	7	58	Trampling	20	50	Conflicting demands	20	53	Inadequate food supply	5	62
Unlawful entry	7	58	Human harrassment	20	50	Degraded scenic viewpoints	18	47	Local attitudes	5	62
Litter	7	58	Exotic plants (V)	19	48	Fire (V)	17	45	Mining	5	62
Inadequate vegetative cover	6	50	Fire (V)	18	45	Trampling	17	45	Inadequate rain (W)	4	50
Exotic animals (V)	6	50				Exotic plants (V)	17	45	Inadequate vegetative cover	4	50
Local attitudes	6	50				Human harrassment	17	45	Inadequate rain (V)	4	50
Conflicting demands	6	50				Unlawful entry	17	45	Trampling	4	50
						Local attitudes	17	45	Loss of habitat	4	50
									Habitat change	4	50
									Conflicting demands	4	50

[a]More than ten threats are listed because of ties.

Note: Threats identified here that were listed under more than one subsystem in the questionnaire have the first letter of the subsystem to which these data apply in parentheses after them.

Table 4.17
Reported Threats for Selected Biome Types (N=number of threats, 1,160)

	Subtropical Forests (n=12)				Tropical Forests (n=23)			
	Suspected		Documented		Suspected		Documented	
Subsystem	N	%	N	%	N	%	N	%
Total	72	30	170	70	206	50	208	50
Water	6	26	17	74	25	49	26	51
Air	0	0	4	100	5	50	5	50
Soil	8	50	8	50	27	63	16	37
Vegetation	8	15	47	85	33	45	40	55
Animal life	23	38	37	62	73	51	69	49
Management	13	25	40	75	32	43	42	57
Other	14	45	17	55	11	52	10	48

Table 4.18
Cause of Reported Threats for Selected Biome Types[a] (N=number of threats, 1,160)

	Subtropical Forests (n=12)						Tropical Forests (n=23)					
	Man		Nature		Both		Man		Nature		Both	
Subsystem	N	%	N	%	N	%	N	%	N	%	N	%
Total	188	78	17	7	34	14	242	58	82	20	71	17
Water	17	74	2	9	3	13	20	39	19	37	10	20
Air	3	75	0	0	1	25	6	60	1	10	3	30
Soil	9	56	1	6	6	38	14	33	13	30	12	28
Vegetation	32	58	7	13	16	29	38	52	17	23	18	25
Animal life	47	78	5	8	6	10	88	62	28	20	23	16
Management	50	94	2	4	1	2	59	80	1	1	4	5
Other	30	97	0	0	1	3	17	81	3	14	1	5

[a]Row percentages by biome type may not equal 100% because of the exclusion of missing and invalid responses.

Warm Deserts (n=9)				Mixed Mountain Systems (n=20)			
Suspected		Documented		Suspected		Documented	
N	%	N	%	N	%	N	%
67	41	95	59	118	34	224	65
4	33	8	67	6	17	29	83
5	62	3	38	5	42	7	58
8	40	12	60	16	41	23	59
10	28	26	72	17	26	49	74
24	51	23	49	38	39	59	61
8	31	18	69	21	33	42	67
8	62	5	38	15	50	15	50

Warm Deserts (n=9)						Mixed Mountain Systems (n=20)					
Man		Nature		Both		Man		Nature		Both	
N	%	N	%	N	%	N	%	N	%	N	%
87	54	33	20	35	22	233	68	22	6	68	20
5	42	3	25	4	33	22	63	2	6	8	23
3	38	1	12	3	38	9	75	0	0	3	25
5	25	6	30	9	45	20	51	2	5	16	41
18	50	7	19	8	22	38	58	5	8	18	27
22	47	14	30	9	19	62	64	13	13	18	19
22	85	2	8	1	4	55	87	0	0	4	6
12	92	0	0	1	8	27	90	0	0	1	3

Table 4.19
Location of Reported Threats for Selected Biome Types[a] (N=number of threats, 1,160)

	Subtropical Forests (n=12)						Tropical Forests (n=23)					
	Inside		Outside		Both		Inside		Outside		Both	
Subsystem	N	%	N	%	N	%	N	%	N	%	N	%
Total	64	26	68	28	105	43	117	28	116	28	150	36
Water	1	4	14	61	7	30	12	24	18	35	16	31
Air	0	0	4	100	0	0	1	10	5	50	4	40
Soil	7	44	3	19	6	38	18	42	2	5	21	49
Vegetation	20	36	6	11	29	53	20	27	16	22	35	48
Animal life	13	22	14	23	33	55	40	28	37	26	55	39
Management	11	21	17	32	21	40	18	24	32	43	12	16
Other	12	39	10	32	9	29	8	38	6	29	7	33

[a]Row percentages by biome type may not equal 100% because of the exclusion of missing and invalid responses.

Table 4.20
Ten Most Reported Threats for Selected Biome Types[a] (N=number of parks reporting, 64)

Subtropical Forests (n=12)			Tropical Forests (n=23)		
Threat	N	%	Threat	N	%
Local Attitudes	11	92	Illegal removal of animal life	21	91
Removal of vegetation	9	75	Fire (A)	18	78
Exotic plants (V)	9	75	Lack of personnel	18	78
Loss of habitat	9	75	Human harrassment	17	74
Conflicting demands	9	75	Fire (V)	15	65
Noise	9	75	Local attitudes	15	65
Fire (V)	8	67	Inadequate rain (A)	14	61
Illegal removal of animal life	8	67	Erosion	13	57
Exotic animals (A)	8	67	Removal of vegetation	13	57
Habitat change	8	67	Unlawful entry	13	57
Lack of personnel	8	67			
Litter	8	67			

[a]More than ten threats are listed because of ties.
Note: Threats identified here that were listed under more than one subsystem in the questionnaire have the first letter of the subsystem to which these data apply in parentheses after them.

	Warm Deserts (n=9)						Mixed Mountain Systems (n=20)					
	Inside		Outside		Both		Inside		Outside		Both	
	N	%	N	%	N	%	N	%	N	%	N	%
	54	33	31	19	62	38	128	37	76	22	116	34
	0	0	4	33	7	58	7	20	12	34	16	46
	1	12	3	38	4	50	1	8	3	25	7	58
	14	70	0	0	5	25	26	67	3	8	10	26
	14	39	4	11	13	36	32	48	13	20	21	32
	10	21	9	19	24	51	31	32	21	22	37	38
	10	38	9	35	5	19	24	38	16	25	12	19
	5	38	2	15	4	31	7	23	8	27	13	43

Warm Deserts (n=9)			Mixed Mountain Systems (n=20)		
Threat	N	%	Threat	N	%
Lack of personnel	8	89	Removal of vegetation	17	85
Illegal removal of animal life	7	78	Illegal removal of animal life	16	80
Erosion	6	67	Erosion	14	70
Trampling	6	67	Lack of personnel	13	65
Loss of habitat	6	67	Trampling	12	60
Compaction	5	56	Loss of habitat	12	60
Inadequate vegetative cover	5	56	Human harrassment	12	60
Removal of vegetation	5	56	Unlawful entry	10	50
Exotic plants (V)	5	56	Conflicting demands	10	50
Fire (A)	5	56	Disease (A)	9	45
Inadequate food supply	5	56	Degraded scenic viewpoints	9	45
Conflicting demands	5	56			
Mining	5	56			

Table 4.21
Cause of Reported Threats by Stage of Economic Development[a] (N=number of threats, 1,611)

	Less-Developed Countries (n=15)						Developing Countries (n=37)						More-Developed Countries (n=46)					
	Man		Nature		Both		Man		Nature		Both		Man		Nature		Both	
Subsystem	N	%	N	%	N	%	N	%	N	%	N	%	N	%	N	%	N	%
Total	148	57	37	14	54	21	351	63	86	15	95	17	552	70	70	9	140	18
Water	6	25	10	42	8	33	30	50	15	25	13	22	52	73	5	7	9	13
Air	2	100	0	0	0	0	5	45	1	9	5	45	20	69	1	3	7	24
Soil	13	46	5	18	8	29	18	30	15	25	25	42	37	51	6	8	29	40
Vegetation	21	43	9	18	17	35	66	59	18	16	22	20	94	57	23	14	45	27
Animal life	56	65	12	14	16	19	104	62	32	19	21	12	132	64	29	14	38	18
Management	39	68	1	2	4	7	95	85	2	2	8	7	131	87	5	3	6	4
Other	11	73	0	0	1	7	33	89	3	8	1	3	86	91	1	1	6	6

[a]Row percentages by development stage may not equal 100% because of the exclusion of missing and invalid responses.

Table 4.22
Location of Reported Threats by Stage of Economic Development[a] (N=number of threats, 1,611)

Subsystem	Less-Developed Countries (n=15)						Developing Countries (n=37)						More-Developed Countries (n=46)					
	Inside		Outside		Both		Inside		Outside		Both		Inside		Outside		Both	
	N	%	N	%	N	%	N	%	N	%	N	%	N	%	N	%	N	%
Total	84	32	42	16	95	36	182	32	134	24	212	38	280	35	211	27	263	33
Water	4	17	4	17	12	50	17	28	21	35	20	33	12	17	37	52	20	28
Air	0	0	1	50	1	50	2	18	3	27	6	55	2	7	15	52	11	38
Soil	15	54	0	0	10	36	24	40	5	8	31	52	54	74	6	8	12	16
Vegetation	17	35	4	8	24	49	42	38	21	19	46	41	71	43	27	16	65	39
Animal life	27	31	14	16	38	44	47	28	40	24	64	38	67	32	47	23	85	41
Management	15	26	17	30	7	12	33	29	39	35	31	28	45	30	50	33	36	24
Other	6	40	2	13	3	20	17	46	5	14	14	38	29	31	29	31	34	36

[a]Row percentages by development stage may not equal 100% because of the exclusion of missing and invalid responses.

Table 4.23
Park Area Affected by Reported Threats by Stage of Economic Development[a] (N=number of threatened subsystems, 490)

	Less-Developed Countries (n=15)								Developing Countries (n=37)							
	No more than 1/4		More than 1/4, Less than 1/2		More than 1/2, Less than 3/4		More than 3/4		No more than 1/4		More than 1/4, Less than 1/2		More than 1/2, Less than 3/4		More than 3/4	
Subsystem	N	%	N	%	N	%	N	%	N	%	N	%	N	%	N	%
Total	29	37	15	19	14	18	11	14	94	54	29	17	19	11	13	7
Water	3	25	1	8	2	17	3	25	16	57	5	18	2	7	5	18
Air	1	100	0	0	0	0	0	0	4	57	0	0	1	14	0	0
Soil	5	45	4	36	1	9	1	9	16	73	1	5	3	14	1	5
Vegetation	8	53	3	20	3	20	1	7	19	63	7	23	3	10	0	0
Animal life	4	27	3	20	4	27	2	13	18	50	7	19	4	11	4	11
Management	4	27	3	20	4	27	4	27	13	39	8	24	6	18	3	9
Other	4	44	1	11	0	0	0	0	8	44	1	6	0	0	0	0

[a]Row percentages by development stage may not equal 100% because of the exclusion of missing and invalid responses.

	More-Developed Countries (n=46)							
	No more than 1/4		More than 1/4, Less than 1/2		More than 1/2, Less than 3/4		More than 3/4	
Subsystem	N	%	N	%	N	%	N	%
Total	117	49	36	15	14	6	43	18
Water	14	50	6	21	4	14	2	7
Air	6	32	0	0	0	0	9	47
Soil	23	72	3	9	1	3	3	9
Vegetation	21	51	11	27	2	5	5	12
Animal life	19	45	9	21	2	5	8	19
Management	19	49	3	8	3	8	11	28
Other	15	41	4	11	2	5	5	14

Table 4.24
Ten Most Reported Threats by Stage of Economic Development[a] (N=number of parks reporting, 98)

Less-Developed Countries (n=15)			Developing Countries (n=37)			More-Developed Countries (n=46)		
Threat	N	%	Threat	N	%	Threat	N	%
Illegal removal of animal life	14	93	Illegal removal of animal life	31	84	Lack of personnel	32	70
Conflicting demands	13	87	Lack of personnel	29	78	Illegal removal of animal life	29	63
Removal of vegetation	12	80	Removal of vegetation	25	68	Erosion	28	61
Fire (V)	11	73	Human harrassment	21	57	Exotic plants (V)	27	59
Lack of personnel	11	73	Erosion	19	51	Trampling	24	52
Erosion	10	67	Fire (V)	18	49	Local attitudes	24	52
Loss of habitat	10	67	Unlawful entry	18	49	Conflicting demands	24	52
Fire (A)	10	67	Local attitudes	18	49	Degraded scenic viewpoints	24	52
Human harrassment	10	67	Fire (A)	17	46	Removal of vegetation	23	50
Unlawful entry	10	67	Loss of habitat	15	41	Exotic animals (V)	22	48
Local attitudes	10	67	Conflicting demands	15	41	Exotic animals (A)	22	48
						Loss of habitat	22	48

[a]More than ten threats are listed because of ties.

Note: Threats that were listed under more than one subsystem in the questionnaire are identified with the first letter of the subsystem to which these data apply in parentheses after them.

Table 4.25
Location of Reported Threats by Affiliation with Special Management Programs[a] (N=number of threats, 1,611)

	Affiliated (n=25)						Nonaffiliated (n=73)					
	Inside		Outside		Both		Inside		Outside		Both	
Subsystem	N	%	N	%	N	%	N	%	N	%	N	%
Total	186	37	104	21	189	37	360	33	283	26	381	34
Water	14	27	16	31	20	39	19	18	46	44	32	31
Air	1	6	7	39	9	50	3	12	12	50	9	38
Soil	39	66	2	3	16	27	54	53	9	9	37	36
Vegetation	44	43	18	17	38	37	86	38	34	15	97	43
Animal life	43	32	24	18	64	47	98	30	77	24	123	38
Management	35	38	22	24	22	24	58	25	84	37	52	23
Other	15	21	15	32	20	43	42	42	21	21	31	31

[a]Row percentages by affiliation may not equal 100% because of the exclusion of missing and invalid responses.

Table 4.26
Park Area Affected by Reported Threats by Affiliation with Special Management Programs[a]
(N=number of threatened subsystems, 490)

Subsystem	Affiliated (n=25)								Nonaffiliated (n=73)							
	No more than 1/4		More than 1/4, Less than 1/2		More than 1/2, Less than 3/4		More than 3/4		No more than 1/4		More than 1/4, Less than 1/2		More than 1/2, Less than 3/4		More than 3/4	
	N	%	N	%	N	%	N	%	N	%	N	%	N	%	N	%
Total	47	34	29	21	16	12	29	21	193	55	51	14	31	9	38	11
Water	6	32	4	21	3	16	4	21	27	55	8	16	5	10	6	12
Air	1	9	0	0	0	0	7	64	10	62	0	0	1	6	2	12
Soil	10	50	6	30	1	5	2	10	34	76	2	4	4	9	3	7
Vegetation	10	43	6	26	4	17	3	13	38	60	15	24	4	6	3	5
Animal life	8	32	6	24	5	20	4	16	33	49	13	19	5	7	10	15
Management	7	32	4	18	2	9	6	27	29	45	10	15	11	17	12	18
Other	5	29	3	18	1	6	3	18	22	47	3	6	1	2	2	4

[a]Row percentages by affiliation may not equal 100% because of the exclusion of missing and invalid responses.

Table 4.27
Ten Most Reported Threats by Affiliation with Special Management Programs
(N=number of parks reporting, 98)

Affiliated (n=25)			Nonaffiliated (n=73)		
Threat	N	%	Threat	N	%
Illegal removal of animal life	20	80	Illegal removal of animal life	54	74
Lack of personnel	19	76	Lack of personnel	53	73
Removal of vegetation	18	72	Removal of vegetation	42	58
Trampling	17	68	Local attitudes	42	58
Erosion	17	68	Erosion	40	55
Exotic plants (V)	14	56	Conflicting demands	40	55
Loss of habitat	14	56	Fire (V)	38	52
Exotic animals (A)	14	56	Human harrassment	37	51
Degraded scenic viewpoints	14	56	Loss of habitat	33	45
Unlawful entry	13	52	Fire (A)	33	45

Note: Threats identified here that were listed under more than one subsystem in the questionnaire have the first letter of the subsystem to which these data apply in parentheses after them.

demonstrates a method of investigating data about threats and suggests several relationships that may merit study.

A chi-square statistic was used to determine if the number of respondents who reported each specific threat listed in the questionnaire differed significantly among parks belonging to different categories of the key study variables. The lambda statistic was used to gauge the strength of the association (for an explanation of these statistics, see Appendix A).

Table 4.28 illustrates the results for parks of different sizes. The four size categories used earlier were replaced by two categories—between 10,000 and 100,000 ha, and between 100,000 and 1,000,000 ha—and only parks in these size categories were included. Four reported threats were found to have statistically significant differences: exotic animals, blocked waterways, loss of habitat, and unsafe conditions. Surprisingly, these threats were reported by a higher percentage of larger parks than smaller ones. For example, loss of animal habitat was reported by 84 percent of the larger parks, compared to 50 percent of the smaller parks. Except for exotic animals, the relationships are extremely weak, as shown by the lambda values.

To examine the relationship between biome type and specific threats, the two most common biome types were used—tropical forests and mixed mountain systems. As can be seen in Table 4.29, only one reported threat differed significantly, although its impact extends to three park subsystems. Inadequate rainfall was reported as a threat to animal life for 78 percent of the tropical forest parks, in contrast to 14 percent of

Table 4.28
Reported Threats Whose Status Significantly Differs by Park Size[a]

Threat	Subsystem	Percentage of Respondents		Significance Level[b]	Lambda Value
		More than 10,000, Less than 100,000 ha	More than 100,000, Less than 1,000,000 ha		
Exotic animals	Animal life	38	74	**	0.30
Blocked waterway	Water	12	36	*	0.00
Loss of habitat	Animal life	50	84	*	0.00
Unsafe conditions	Management	12	43	*	0.00

[a] Includes parks more than 10,000 and less than 1,000,000 ha in size
[b] *Significant at $p<.05$; **Significant at $p<.01$; ***Significant at $p<.001$

Table 4.29
Reported Threats Whose Status Significantly Differs by Biome Type[a]

		Percentage of Respondents			
Threat	Subsystem	Tropical Forest	Mixed Mountain Systems	Significance Level[b]	Lambda Value
Inadequate rain	Animal life	78	14	***	0.63
Inadequate rain	Vegetation	67	15	**	0.43
Inadequate rain	Water	67	17	***	0.40

[a]Includes parks from tropical forest and mixed mountain system biomes
[b]* Significant at $p<.05$; **Significant at $p<.01$; ***Significant at $p<.001$

those in mixed mountain systems. The lambda values suggest a moderately strong relationship.

In contrast, a great number of specific threats differed significantly by stage of economic development. Table 4.30 shows that the number of respondents who reported 14 specific threats was significantly different among parks in countries at different stages of economic development. Every subsystem except soil is represented. Several specific threats were reported more often by parks in more-developed nations: exotic plants, chemical pollution, legal removal of animals, noise pollution, and mining. For some of these threats, the trend is very significant. For example, although none of the parks in less-developed nations reported chemical pollution as a threat, more than one-third of the parks in developed nations so reported. Other threats were more often reported by parks in less-developed nations—including unlawful entry, fire, inadequate rain, harassment of animals, illegal removal of animal life, removal of vegetation, and conflicting demands. The threats with high lambda values were unlawful entry (associated with less-developed nations) and exotic plants (associated with more-developed nations).

Finally, as shown in Table 4.31, four threats significantly differed in their reporting rates for affiliated parks as opposed to unaffiliated parks: exotic animals, overpopulation of animal species, chemical pollution, and trampling. Except for the overpopulation threat, the affiliated parks had higher reporting rates, though differences as measured by the lambda values are fairly weak.

One limitation of this kind of analysis is that each of the key variables is examined separately without the others being controlled. Results will often vary significantly for two or more variables, making it difficult to judge which one might be responsible for the variation. For example, the reported threat of exotic animals varied significantly for both the size and affiliation variables: Larger parks and those affiliated with special programs reported the threat more often than smaller and/or nonaffiliated parks. Likewise, inadequate rainfall (threatening animal, vegetation, and water subsystems) varied significantly for both the biome and economic development variables. Which variable is most influential—size or affiliation, biome type or stage of development? From the data for these threats, it is difficult to tell.

Yet of the 21 reported threats found to vary significantly by the study variables, 17 were associated with only one variable and 10 of those were linked to economic development. Further, four of the threats associated with economic development were in the "top 10" threats listed earlier: human harassment of animal life, conflicting demands, illegal removal of animal life, and removal of vegetation. These four threats were reported by all the sampled parks in less-developed nations.

Table 4.30
Reported Threats Whose Status Significantly Differs by Stage of Economic Development

Threat	Subsystem	Percentage of Respondents			Significance Level[a]	Lambda Value
		Less Developed	Developing	More Developed		
Unlawful entry	Management	77	62	28	***	0.36
Exotic plants	Vegetation	20	46	66	*	0.22
Fire	Animal life	83	61	44	*	0.13
Inadequate rain	Vegetation	58	30	11	**	0.11
Inadequate rain	Water	54	42	8	***	0.05
Human harrassment	Animal life	100	72	49	**	0.04
Chemical pollution	Air	0	9	36	***	0.00
Removal of vegetation	Vegetation	100	71	58	*	0.00
Legal removal	Animal life	0	15	38	*	0.00
Illegal removal	Animal life	100	89	69	*	0.00
Inadequate rain	Animal life	58	46	9	***	0.09
Conflicting demands	Management	100	56	56	**	0.00
Noise pollution	Other	17	6	36	**	0.00
Mining	Other	38	19	46	*	0.00

[a] *Significant at $p<.05$; **Significant at $p<.01$; ***Significant at $p<.001$

Table 4.31
Reported Threats Whose Status Significantly Differs by Affiliation with Special Management Programs

Threat	Subsystem	Percentage of Respondents		Significance Level[a]	Lambda Value
		Affiliated Reported	Nonaffiliated Reported		
Exotic animals	Animal life	70	40	*	0.22
Overpopulation	Animal life	28	65	*	0.20
Chemical pollution	Water	50	23	*	0.00
Trampling	Vegetation	81	50	*	0.00

[a]*Significant at $p<.05$; **Significant at $p<.01$; ***Significant at $p<.001$

NOTES

1. This response rate is based on a sample of 129 parks since six unopened survey packages were returned because of inadequate addresses. In addition, two surveys were returned too late to be included in the data analysis. Hence the response rate for usable surveys was 73 percent.

2. Sixteen percent of the potential responses were invalid or missing.

CHAPTER 5

Peril, Opportunity, and Choice

The Third World Congress on National Parks was held in Bali, Indonesia, during fall 1982. One of the early speeches at this meeting was made by Peter S. Thacher, executive director of the UN Environment Programme. He forthrightly laid out the challenges facing *Homo sapiens* in the last years of the twentieth century:

> The *peril* we face is the accelerating destruction of the living resources of this planet in a way which strips diversity, increases risks of instabilities, and undercuts the basis for sustainable development. The *opportunity* is to change course and carry out the World Conservation Strategy. The *choice* is whether we pay now or later. (McNeely and Miller 1984:12)

The data presented in Chapter 4 provide an emerging profile of the perils parks face around the world. The first purpose of this chapter is to suggest some general conclusions derived from the data. Our conclusions are somewhat cautious, for the study's limitations must be remembered: The information is based on respondents' perceptions of threats; only a modest sample of national parks was included; the survey ignored the important topic of cultural resources. Nevertheless, there may be value in stating, as simply as possible, what we think the data mean.

Second, we make several recommendations for dealing with threats to parks. Obviously, practical solutions to the problems that beset national parks are not easy to find. Wendell Berry's agricultural wisdom applies to our efforts:

> Good solutions exist only in proof, and are not to be expected from absentee owners or absentee experts. Problems must be solved in work

> and in place, with particular knowledge, fidelity, and care, by people who will suffer the consequences of their mistakes. Practical advice or direction from people who have no practice may have some value, but its value is questionable and is limited. (1981:134)

Nevertheless, we have made our suggestions as concrete as possible, for we see opportunities to deal effectively with many threats to national parks, not just in theory, but in the field. A historic example of such an opportunity took place in Olympic National Park in the United States where, during World War II, pressures were intense to log Sitka spruce for airplane manufacturing. After careful fact finding and negotiation, other sources of wood were found, and the Olympic Sitka spruce preserved. Joseph Sax notes: "The Olympic Park case reveals that claimed conflicts are often less intractable than they appear at first view; that by forcing alternatives explicitly into the open and by pursuing the facts behind the claims, we can often resolve concrete cases without having to weigh competing values in the abstract" (1982:61).

Our recommendations are directed to international conservation and development organizations, to park agencies and park managers in the field, and to scientists conducting research related to threats. We end with a brief but strongly held opinion concerning the choices that will affect the future of national parks.

CONCLUSIONS FROM THE DATA

1. Substantial and diverse threats confront the natural resources of national parks worldwide.

The national parks of the world currently face threats to every subsystem—air, water, soil, vegetation, animal life, and management. Every potential threat listed in the questionnaire was reported by personnel in at least one park. Without time series data, it is admittedly difficult to estimate whether these problems are increasing or decreasing. Yet in this survey, more than 1,600 threats were reported. More important, only three managers reported no threats to their parks.

2. Many threats are not well documented.

Although the problems facing national parks are substantial and diverse, our knowledge of them is often inconclusive. A full 40 percent of the reported threats were suspected rather than documented. Documentation was lowest for the soil, animal life, and air subsystems.

3. Although man may be the principal cause of threats to parks, Nature plays an important role as well.

Man was implicated as at least a partial cause for a large majority of the threats; yet Nature's role as a threat to parks was not inconsequential, as paradoxical as that sounds. Twelve percent of the reported threats were attributed solely to Nature, and that percentage increased for certain subsystems in certain biomes. Further, the data illustrate the interplay of man and Nature in producing threats to parks; one-fifth of the total reported threats were attributed to *both* man and Nature.

4. Threats are located both within and outside park boundaries.

One-third of the reported threats were reported as internal problems. Mining, for example, was reported by 32 percent of the sample; 46 percent of the parks in industrialized nations so reported. An additional one-third of the threats were cited as being located both inside and beyond park boundaries. The sources of many other reported threats were identified as outside parks. Underscoring the theoretical framework in Chapter 3, several of the 10 most reported threats in the study (illegal removal of animal life, removal of vegetation, local attitudes, and conflicting demands) can be attributed to interactions between parks and socioeconomic regions adjacent to them.

5. Most reported threats are localized within park areas.

Most reported threats to subsystems studywide were found to have localized effects on park areas. Similar results were obtained when responses were analyzed by subsystem. Reported threats to the air subsystem were most frequently cited as having generalized effects on park environments; still, only one-third of these threats were perceived to affect more than three-fourths of the park.

6. Animal life, management, and vegetation subsystems are most often affected.

Threats to these subsystems accounted for most of the reported threats studywide. Nine of the 10 most frequently cited threats in the study potentially affect one of these subsystems. It is especially noteworthy that threats to management are so highly ranked; this suggests additional problems in dealing with threats to other subsystems.

7. Many threats are common to parks worldwide, but the frequency of their occurrence differs.

The data suggest a core of threats, perhaps linked, that threaten many national parks. Nine threats were reported by a majority of the managers; illegal removal of animal life and lack of personnel were the most frequently cited. By subsystem and by the key study variables, many threats were consistently reported, but their ranks differed.

8. The characteristics of reported threats differ (sometimes significantly) among parks by size, biome type, stage of economic development, and management program.

The data on park size and biome type suggest some differences, but the low number of comparisons (only a few biomes were compared, for example) mean our results are inconclusive. Of the four study variables, economic development emerged as clearly influential. Numerous threats varied systematically among parks in less-developed, developing, and more-developed countries. The core of threats just described may vary at each stage; in medical terms there may be a "threats syndrome" typical to parks in countries at each stage of economic development.

The Assumptions Reexamined

Another way to interpret the data is to juxtapose the results with the general assumptions we made at the beginning of the book. Do the conclusions support or contradict our earlier statements? We first assumed that biosocial regularities lead to predictable and common behaviors in and around parks. The 10 most reported threats, the core of threats, seem to reflect such a pattern—poaching, removal of vegetation, conflicting demands all threaten a majority of national parks; yet they are far from universal. Second, we assumed that cultural and environmental variation results in many diverse and culturally specific behaviors. The wide variation in status of certain threats (such as chemical pollution) and the long list of written-in threats (from coffee pulp dumping to volcanic activity) support this assumption.

Third, we assumed that *Homo sapiens* is ecologically interdependent with the natural world and that man-Nature relations are causal and bilateral. Almost all our data—the reporting of so many threats, the perceived causes being humans, Nature, or both, and the descriptive accounts provided by managers—support this assumption as it applies to parks. Not documented by our study, however, is the influence of Nature upon social systems.

Our fourth assumption was that man-Nature relations can best be understood using a systems perspective. We think that the data reflect the usefulness of systems thinking and that by examining threats to park subsystems rather than parks in general, advances can be made in monitoring, mitigating, and managing threats.

The Study Variables Reexamined

Still another way to interpret the data is to evaluate the importance of the key study variables—size, biome type, stage of economic development, and affiliation with special management programs. Our data associating threats and size of reserve are mixed; only a few threats significantly varied by size, according to statistical analysis, though they included such important concerns as loss of habitat and exotic animals. Although some variation was found at the subsystem level, threats reported to parks of different sizes were similar in status, cause, location, and area affected.

Results for the biome variable were even more inconclusive. Most of the threats to parks in each biome type were reported to be documented and caused by man, and to have generalized effects on park environments. Some variation occurred at the subsystem level; yet only one threat (inadequate rain) was statistically associated with biome type when tropical forest and mixed mountain parks were examined. The lack of comparative data for all biome types makes premature any kind of conclusion about the usefulness of biome type as a variable.

Unlike biome type, stage of economic development emerged as a potentially powerful variable. For example

- Parks in more-developed nations reported the highest percentage of documented threats.
- The percentage of threats attributed to man increased from developing to more-developed countries; threats caused by nature decreased slightly from developing to more-developed nations.
- Outside sources of threats were most frequently reported by parks in developing and more-developed countries.
- Reported threats to subsystems had localized effects for a majority of the parks.
- The most numerous threats to parks in developing and developed countries were identical, but most were ranked differently.

In addition, numerous threats showed statistically different reporting rates for parks in countries at different stages of development, even at high significance levels of .01 or .001. Stage of economic development is clearly an important variable influencing threats to parks.

Finally, our data on affiliation with MAB and the World Heritage Program suggest that affiliation may have minimal association with the reporting of threats. The status, cause, and location of reported threats in affiliated and nonaffiliated parks were found to be similar. The most reported threats for affiliated and nonaffiliated parks were identical, and few threats significantly varied by this variable.

THE ROLE OF INTERNATIONAL ORGANIZATIONS IN MANAGING THREATS TO PARKS

Current trends in population growth and resource consumption, combined with our data, suggest that parks in less-developed countries will be among the most intensely pressured in the future. Yet their resources to meet these problems are minimal, and the role of international organizations is likely to increase.

Threatened parks are found in the more-developed nations as well, but these countries simply have more resources—money, skilled manpower, organization—and hence more options. Therefore, priority might be given to assisting less-developed countries in building the infrastructure necessary to sustain parks—politically, economically, socially, and biologically. A systems approach may be needed; preservation strategies directed at a single species or subsystem may in the long run be insufficient and ineffective. Actions might include

1. Providing support for nongovernmental conservation groups that can aid efforts to mitigate and prevent threats
2. Providing support for the development and enforcement of national and local environmental laws to help mitigate threats
3. Continuing support for programs such as the World Conservation Strategy to help mitigate and prevent threats that result from local socioeconomic development
4. Increasing training of personnel on topics related to threat management and providing funding for parks in developing regions specifically to help train resource management specialists

In addition, national parks, "protection forestry," Nature-based tourism, and other conservationist land uses should be seriously explored as real alternatives in economic development plans. Elsewhere we have suggested why several of these strategies receive little attention from traditional development organizations and resource managers:

> Protection forestry (which is primarily the management of forest lands in parks and reserves) is fundamentally different from other forms of forestry. Economic exchanges are not derived from the cash value of the timber, nuts, twigs, or total biomass, but from the social characteristics of the forest ecosystem—its attractiveness, uniqueness, beauty and so forth. The sale of one million board feet to a multinational corporation is a comfortable transaction to the forester; the sale of 1,000 package tours to a travel agent is not. This may reflect trained incapacities rather than economic

> reality. Both transactions may provide economic benefits and incur social and environmental costs; both demand careful forest management. (Machlis 1984:51–52)

Development of lands adjacent to parks should be carefully considered. In underdeveloped nations, intensive development, particularly agricultural, might actually reduce threats by providing local foodstuffs, building materials, and wood for fuel. Such development might need to satisfy a range of specialized criteria: sustained yield, low level of wastes, and minimal risk to locals, with a balance between sustenance and cash income. An archipelago pattern of ecosystem preservation in nearby areas could be tried, involving islands of long-rotation management, carefully chosen by design rather than default, as Harris (1984) suggests. The limits may have to be set around fertilizer and pesticide dispersal patterns, and the yields may need to be adjusted as cash prices and per capita needs change, but such problems seem tractable. One manager writes:

> Deriving this balance, of course, is probably the most difficult job for the park managers, but we can make headway with increased research effort. . . . [We] need to arrive at quantities of timber, animals, etc., that can be cropped for rural use without impacting the ecological balance of the park ecosystems. Lack of research personnel and funds is therefore a compounding problem.

Such interdisciplinary research is clearly beyond the resources of most national parks or even national park agencies; international initiative is necessary.

Our data revealed that a core of threats often exists in parks, regardless of which biome type they encompass, what economic characteristics exist in the country in which they are found, and whether parks are affiliated with special management programs. The 10 most reported threats and the three most threatened subsystems (vegetation, animal life, and management) might become the focus of international efforts to mitigate and prevent threats. Actions might include

1. Using the lists of threats in this book in budgeting, planning, and setting priorities for various aid projects.
2. Establishing a coordinated, large-scale, and worldwide program to deal with poaching—the most widespread threat to parks. Poaching was reported as a threat to wildlife in 76 percent of the parks surveyed: 93 percent of the parks in less-developed nations and 84 percent of those parks in developing nations. Although trophy

poaching by organized crime is undoubtedly important, such a program must deal effectively with local poaching for subsistence as well. A game warden writes: "The majority of the people are hungry and lack reliable and wholesome food supplies, particularly animal protein. . . . In most areas in the country 'bushmeat' is considered as a luxury. Poaching, therefore, is the number one threat to animal life in national parks, game reserves and other conservation areas."

As the survey data showed, poaching is not a problem associated only with underdevelopment; it was the second highest ranked threat in more-developed countries, with 63 percent of the parks in these industrialized nations reporting illegal removal of animal life. Consequently, any worldwide antipoaching effort must extend to populations with relatively high per capita incomes, substantial diets, and alternative food sources. Why do locals poach? How do they go about it? What are the consequences for park ecosystems? How can poaching be reduced? These all are crucial questions that could be addressed by a worldwide integrated program of training, research, enforcement, and management.

3. Developing a model "threats management program" linked with demonstration projects in countries at various stages of economic development to experiment with practical solutions to the core threats faced by most national parks.

Just as the concept of national parks must be adapted to meet regional conditions, the idea of coordinated management of threats to parks must be customized as well. Our data revealed variations in the status, cause, location, and extent of threats among parks in different regions of the world. Therefore, great care needs to be taken to find individually designed solutions to common problems. Threats must be dealt with, not by national policies established on paper, but by on-the-ground striving for improvements—reduced industrial pollution, more manageable visitation, increased park staff, and so forth. The phrase "think globally, act locally" applies; international groups can continue to organize their efforts on a regional basis. Actions might include

1. Holding periodic workshops, training sessions, and conferences, all involving field managers and aimed at providing technical assistance to parks that share similar biome types or stage of economic development. Experts from fields outside of traditional park management should be included, for it may be a weakness that park people are training park managers. Agricultural econ-

omists, rural development specialists, foresters, conservation biologists, and others can greatly contribute to such training programs.
2. Expanding communication, including new field-level networks, among managers of similar parks with similar problems and between park managers and other resource managers.
3. Improving accessibility to scientific data relevant to threats. With so many of the reported threats suspected, managers need access to scientific information as a precursor to decision making. Yet few parks will be able to document fully the threats they face for want of time, personnel, and funds. Hence a central data base for field managers is a crucial diagnostic tool.
4. Dispersing research and educational opportunities worldwide. These would include additional financial support for the development of interdisciplinary park management curricula at universities in less-developed countries. These curricula, and those in the United States and Europe, should include courses in management of threats.

STRATEGIES FOR NATIONAL AGENCIES AND PARKS

Although international organizations may play an important role in dealing with threats, national parks are managed by sovereign nations with unique political, economic, cultural, and environmental histories. National strategies are crucial for managing threats to national parks. In some cases, new organizational frameworks might be experimented with. For example, the widely accepted goal of autonomy for national park agencies may not always be beneficial because interagency rivalry and communications problems can hinder wise management. Since in many countries, agricultural and development policies often supersede conservationist objectives, integrating park management into these departments may improve the status and governmental awareness of national parks. At least autonomy for conservation organizations does not need to be accepted as dogma.

In addition to organizational changes, policy alternatives may need to be considered. For nations with large and diverse park systems, threats pose problems in administering national policies while at the same time dealing with local situations. Therefore, policies combining strong objectives with flexibility in meeting these objectives (say, an air quality standard) are advantageous.

Many threats are reported to originate at least partially from human activities on lands adjacent to parks. Pressures on these buffer areas will increase as will their impact upon park systems. Therefore, techniques for mitigating threats from buffer areas should be aggressively explored. Actions might include

1. Zoning adjacent lands for compatible uses
2. Involving local people in park planning and management
3. Compensating local communities for loss of resource base, perhaps by taxing nonlocal visitors
4. Conducting public education programs to create local awareness of park values
5. Purchasing development rights
6. Developing adjacent lands to expand regional resource base outside park boundaries
7. Controlled cropping of park resources

All these suggestions have been put forward by other authors, and some have been applied in one national park or another. J. A. McNeely and K. R. Miller provide numerous case studies in *National parks, conservation and development* (1984). We suggest that the key effort should be to evaluate how (or if) these programs reduce threats to parks and to develop national programs for adopting the most successful approaches.

Finally, lack of personnel was identified as a threat by 73 percent of the parks worldwide. This threat to effective management may be linked to numerous other threats, for personnel are needed to solve most park problems. Therefore, national governments, park agencies, and local superintendents need to make every effort to increase the number of resource managers in the field. Actions might include

1. Increasing funding for park rangers and resource management specialists, particularly for parks faced with threats that require personnel (poaching is an example)
2. Educating visitors and local populations about park values and regulations, so that personnel can be free to complete tasks unrelated to law enforcement
3. Renewing interest in using workers from development and aid agencies like the U.S. Peace Corps to bolster staff in critical areas

CRITICAL RESEARCH NEEDS

Several critical research needs are associated with threats to parks. First, our description of reported threats to parks provides a potential research agenda for natural and social scientists alike. The core of common threats, combined with their low level of documentation, call out for applied research activities. Research on threats to parks should be encouraged, especially for the core threats. Actions might include

1. Setting research agenda for parks in each biome type, as well as an international agenda dealing with the core threats
2. Providing research funds earmarked for research relevant to threats
3. Establishing a network of scientists conducting research on various park subsystems, particularly those most threatened—animal life, vegetation, and management
4. Convening an international symposium on research in national parks
5. Encouraging applied preservation sciences, such as protection forestry, wildlife reproduction, and landscape ecology, including expanded course offerings at the university and professional levels

Currently such fields of study are held apart by the orthodoxies of academic organization. L. E. Gilbert states:

> Unfortunately, the fragmented structure of our academic, research, and funding institutions tends to isolate the disciplines of basic ecology, agriculture, conservation and environmental protection. At best, there is little overlap between these areas; at worst, there is active antagonism. It is incumbent upon those concerned with the long-term biological diversity of the earth to ignore the traditional boundaries of scientific subdisciplines and to seek, instead, more holistic solutions. (1980:32)

Although the exact configuration of a "preservation science" is unclear, its unifying concern could be the resilience of ecosystems.

Second, although our study provides a profile of threats to parks, it does not examine the dynamic nature of threats in any detail. Yet detailed knowledge of how threats occur, are perceived, and influence park ecosystems is crucial. Research might be focused on better understanding the ecology of threats. The emphasis of such research should be upon applied studies, with experimentation, monitoring, and evaluation being important research techniques. S. I. Auerbach states: "The aim of sound ecological policy is not to predict and eliminate future surprises, but rather to design resilient systems which can absorb, survive, and capitalize on unexpected events when they do occur" (1981:29). Actions might include

1. Systematically investigating which threats are linked together and how these linkages change as regional development proceeds. An effort to correlate certain specific land uses and specific threats

would be valuable, for such studies could help develop effective mixed land-use plans and policies.
2. Studying in detail specific threats to identify their pathways and consequences to park subsystems, parks, and regions.
3. Studying carefully the consequences of current threats (as natural experiments) to develop knowledge of ecosystem resiliency. Comparative and longitudinal studies will be especially valuable.

Third, the results of this study suggest that all around the world, activities on lands adjacent to parks influence the threats that occur. However, little research has focused on the human wants and needs that partly perpetuate these activities. Research should be conducted on the local populations living near park boundaries. Numerous questions could be considered:

- What are the wants and needs of local peoples? Can their resource demands be reasonably predicted?
- How do local attitudes become translated into activities that support or threaten park objectives?
- What benefits do local populations actually receive from national parks?

Fourth, our survey of threats to parks can be improved upon in numerous ways. The worldwide scope required use of general categories to operationalize the key variables of biome type and stage of economic development. A future study might be confined to one continent so that more detailed biogeographic and economic data could be used to operationalize the variables and hence provide more specific results.

Our effort ignored threats to cultural resources, such as buildings, roads, historic sites, and trails. However, these resources are critical elements of many national parks. The widespread nature of such threats as lack of personnel, erosion, local attitudes, and fire suggests that the cultural resources of the world's parks are probably not untouched; threats to the cultural resources of national parks worldwide should be investigated as soon as possible.

Statistical analyses were limited by the relatively small number of parks included in our sample. Future studies might include a larger sample or even a complete census that would facilitate responses from more parks and the use of additional statistical tests to investigate relationships among threats.

Finally, this study provides a cross-sectional view of the state of the world's parks. Long-term patterns and trends are difficult to derive from this type of analysis. A survey of threats to the world's parks should be conducted periodically (say, every five years) to provide longitudinal data from which trends in threats could be determined.

A CONCLUDING OPINION

This study demonstrates that park managers around the world perceive significant and wide-ranging threats to the natural resources of national parks. It is apparent, however, that the implications of these findings extend beyond the confines of professional park management. A difficult question arising from the analysis of threats to parks is how trade-offs among preservation, conservation, and development should be handled. How should these choices be made?

In our opinion, the debate will be political rather than scientific, and the subject matter will only be what trade-offs will most benefit *Homo sapiens*. Other species and ecosystems will have little standing, other than as foils for human needs and wants. George M. Wright, the scientist who studied the U.S. national parks, noted in 1933:

> How shall man and beast be reconciled in the conflicts and disturbances which inevitably arise when both occupy the same general area concurrently? As man is at once poser of the question, arbiter in the arguments, and, above all, himself the executioner, his verdict will be determined directly by the use he wants to make of any particular area and the order in importance to him of those uses. (Quoted in Wauer and Supernaugh 1983)

We must always remember that national parks, for all their seeming wildness and the apparent dominance of Nature are partly social creations. They are conceived, established, maintained, and in turn threatened by society. Lost in the philosophical debates between "environmental protection" and "resource development" is the fact that the perpetuation of national parks does not depend upon one or the other; it depends on both.

The economic and environmental decisions necessary for the survival of national parks will in our opinion be based on who wields political power. As mentioned before, in the more-developed nations the trade-offs are especially complex, for the alternatives and resources are so relatively plentiful. The pressures placed upon parks by mining, forestry, commerce, agriculture, and tourism often can reduce the management of threats to a series of political compromises, incrementally protecting or harming park ecosystems. In the developing nations of the world, the choices are even harsher. As D. Mueller-Dombois et al. state: "The practical realities of developing nations are that exploitation of natural systems must occur so that the human condition can be improved" (1983:6). Stuart Marks, writing about East Africa, illuminates the fewer alternatives with pessimism: "The romantic vision of keeping Africa as an unchanged paradise teeming with wildlife is foreign nonsense, for to ask East Africa to perpetuate such an image is to ask it to stay poor and underdeveloped" (1984:135).

Yet the romantic vision of parks as protected paradises is widespread and, ironically, may threaten the permanence of national parks. This purely preservationist approach, where parks are considered "fortresses" under siege, invincible or soon eradicated, carries great political risks. It requires an essentially militaristic defense strategy and will almost always heighten conflict. One manager, writing on the back of his questionnaire, commented:

> There is war in our national park system, because if a poacher sees you first then you will be shot on sight. The best thing to do I guess is to devise a system whereby departmental law enforcement officers should be armed with automatic firearms to meet the challenge from poachers.

Yet another wildlife research officer provided a more long-range insight:

> The administration does not have an adequate law enforcement staff component to counter illegal off-take and destruction of the park's resources. With the existing staff and law enforcement effort, however, I see indications of overt conflict between park personnel and rural populations. I therefore think that an increase in law enforcement effort, even though desirable, may increase the conflict, if present park objectives are maintained. A new policy, to include a more flexible approach to the needs of rural populations, is required.

Parks as fortresses have other risks, especially where the elite of a country controls access and use. For example, the concept of social carrying capacity, so attractive to contemporary managers in the United States, may prove as exclusionary in developing nations as developed ones (see Burch [1981] for a thorough discussion).

In contrast to the fortress mentality is the human ecological view espoused in this book. The central theme is long-term productivity or sustainability of park systems. Long-term stability may be more valuable than sporadic, extreme preservation, which often cannot be maintained. R. Allen stresses:

> Development is essential for conservation, because without it conservation cannot be sustainable either. In many developing countries especially, where plants, animals and their habitats are being destroyed primarily by a combination of poverty and population growth, the surest and most effective way of conserving those resources is probably a set of simple development measures: Establishment of fuelwood plantations; provision of appropriate technologies (such as more efficient stoves); incentives and training to convert from shifting to settled agriculture; a direct stake in revenues from tourism; and so on. Unless such measures are taken, it is likely that traditional conservation measures (such as the establishment of protected areas and the strengthening of anti-poaching forces) will fail rather quickly or, worse, backfire. (1980:5)

Allen's comments may also apply to parks in rural areas of industrialized nations. Lobbyists for environmental organizations might better spend their time lobbying for research programs, "soft" technologies, and energy conservation efforts than the total exclusion of all energy development from lands adjacent to park boundaries. This argument should not be confused with support for simple economic exploitation; clear-cutting of forests, large-scale hydroelectric projects, and surface mining are relatively crude resource practices inappropriate for national parks.

We suggest that parks be seen as the multiple-use areas they already are—providing watershed protection, recreation, tourism expenditures, employment, and so forth. Their integration into regional socioeconomic systems is, we believe, the crucial element that will sustain them in the long term, regardless of a nation's stage of development. Some measurable benefits must flow from park to region. When a threat to a national park is perceived by the regional population as a threat to its own well-being, such integration will be complete, and the wise husbandry of parks more readily accomplished. N. Myers warns:

> A national park is as integral to its regional environment as it is to the nation. Whether one wants to manage it that way or not, a park is dependent on the resources—human and physiobiological—of the environs, just as the environs are modified by the park's existence. The ramifications of this relationship, especially the socioeconomic ones, are not always recognized, with the result that the enmeshing process sometimes sounds like a crunching of the gears. The worlds on both sides of the park boundary would get along better if there were a clear indication of what each can do for the other. By contrast, if they spend their energy resisting

> one another, there is little doubt as to which must be the ultimate "winner." (1972:1262)

And even the "winner" would ultimately lose—in the inevitable depletion of watershed, plant and wildlife species, and the wonder of Nature's processes. They may not be paradise, but a world without national parks would be a poorer place, indeed.

APPENDIX A

Notes on Research Methodology

OPERATIONAL DEFINITIONS

The size of individual national parks in hectares was obtained from the IUCN's *1982 United Nations list of national parks and equivalent reserves* (IUCN 1982b). Parks were placed in one of four size categories: (1) less than 10,000 ha; (2) 10,000 or more, but less than 100,000 ha; (3) 100,000 or more, but less than 1 million ha; and (4) 1 million or more, but less than 10 million ha. These divisions were adapted from a similar categorization of units in the National Park Service in the United States (Schonewald-Cox 1983).

The biome types for parks included in the study were obtained from the 1982 UN list of national parks and protected areas (IUCN 1982b) where parks are assigned a biome type based on this classification.[1]

The World Bank data on labor force by sector were used as an indicator of economic development. The labor force encompassed economically active persons including the armed forces and the unemployed but excluding housewives, students, and economically inactive groups. The agricultural sector is composed of agriculture, hunting, and forestry activities (World Bank 1981).

A list of parks included in the Man and the Biosphere and World Heritage Trust programs was obtained from D. Hinrichsen (1983) and the IUCN (1982a). Parks involved in one or both of the programs were included in the sample and identified as "affiliated" parks.

CONSTRUCTION OF THE SAMPLING FRAME

The sampling frame consisted of a list of all national parks derived from the *United Nations list of national parks and equivalent reserves* (IUCN 1982b).

Each park in the sampling frame also had to have an adequate local mailing address so that managers at the park site could be surveyed. A variety of sources, including *The world directory of national parks and other protected areas* (IUCN 1975) and the IUCN's Protected Areas Data Unit (PADU) in the United Kingdom, were consulted to obtain these addresses. However, complete local addresses were unavailable for many parks. If a park was important to the study design (for example, if it was the only park within a particular biome type) and a regional or national office address rather than a local one existed, it was used. Moreover, few local addresses were available for parks in several countries; for these other nonlocal addresses were employed, particularly if park staff such as the superintendent appeared to be located at this address.

A limited number of parks smaller than 10,000 ha, with few personnel or located in countries such as Iran, which had poor diplomatic ties with the United States, were included in the sampling frame to decrease the probability of nonresponse. For the same reason, not more than one park was surveyed at any given address. The sampling frame consisted of 318 national parks and equivalent reserves.

SELECTION OF THE SAMPLE

Parks in the sampling frame were stratified according to four key variables: biome type, stage of economic growth, management program affiliation, and size. The first criterion was to choose parks that would provide a representative distribution by biome type (see discussion in Chapter 4). Another criterion was to obtain similar numbers of parks from countries at each of the three stages of economic growth so that the threats to parks in countries at different stages of economic development could be compared. The sample chosen consisted of 23 parks in less-developed countries, 56 in developing countries, and 56 in more-developed countries.

The third criterion was to include as many parks as possible that already met the first criterion and were affiliated with either the Man and the Biosphere or World Heritage Trust programs. This requirement provided an opportunity to compare threats between parks in these programs and those that were not affiliated. Only world heritage sites nominated for their natural significance were included in this group since the study's emphasis was on natural rather than cultural resources. The final sample consisted of 33 affiliated parks (18 biosphere reserves, 8 world heritage sites, and 7 parks affiliated with both programs) and 102 nonaffiliated parks.

The final criterion was to choose a diverse sample of parks with regard to size; parks in the sample ranged from 1,000 to 70 million ha.

Appendix B lists the categorization of each park in the sample according to the four key variables.

Questionnaire Design and Construction

Although the feasibility of conducting international mail surveys is improving (that is, mail service and literacy rates have advanced in many areas), research necessary for attaining better quality, reliability, and response rates is lacking. The literature on international mail surveying includes general treatments in textbooks (Erdos and Morgan 1970) and case studies that focus on particular techniques that may improve response rates such as personalization and monetary incentives (see, for example, Eisinger et al. 1974).

Most information on international surveying must be derived from experience with personal interviews. This information is generally limited to problems associated with linguistic and conceptual equivalence and cultural response bias (Almond and Verba 1963; Sicinski 1970; Warwick and Osherson 1973).

Because of the lack of information on international mail surveying, methods of questionnaire construction and implementation in this study were adapted from D. A. Dillman's total design method (1978), which has been shown to increase response rates in the United States. In comparison to other literature on mail surveying, Dillman (1978) provides a more theoretically based and comprehensive discussion of survey design, layout, and implementation. The subsystems used in the questionnaire were defined as follows:

1. *Water* is found in different natural forms, such as lakes, rivers, and oceans. Only ground and surface water found within park boundaries is included in this subsystem.
2. *Air* refers to the particles and layers of gases that surround the earth and make up its atmosphere.
3. *Soil* refers to the loose mineral and organic material on the surface of the earth in which plants grow.
4. *Vegetation* refers to living photosynthetic organisms (plants) such as trees and grasses, which can manufacture their own food from inorganic substances.
5. *Animal life* refers to living nonphotosynthetic organisms (animals), such as birds, mammals, and fish, which are unable to manufacture their own food from inorganic substances.
6. *Management and administration* refers to activities such as development and protection, which are required to manage and administer all park resources as a whole.

7. The *Other* category refers to threats that are not specific to any of the resource categories previously mentioned.

Each subsystem was included with a list of potential threats that applied specifically to it. The lists of threats were compiled from literature cited in Chapter 3. Park managers were first asked whether a particular resource subsystem (with the exception of the "other" category) was threatened in that park. If it was, managers were asked three questions concerning the status, source, and location of each threat listed. They were also asked to indicate how much of the park was affected by threats to each subsystem. In addition, blank spaces were provided to enable respondents to write in threats not listed. Questions concerning the attributes of the respondents such as the level of education and years of experience also were included. In most cases the questions were close ended to minimize translation and coding problems.

Pretests

Several different pretests were conducted to determine whether wording or methodological problems existed in the questionnaire. An abbreviated form of the questionnaire was used to survey participants at the World National Parks Congress in Bali, Indonesia. Drafts of the questionnaire also were sent to approximately 20 professionals in natural resources and social science fields to solicit their suggestions and comments on ways to improve it. In addition, a draft of the final questionnaire was reviewed by several international students at the University of Idaho and a visiting class of U.S. Agency for International Development (USAID) participants from 14 countries. Observations and suggestions from these pretests were incorporated into the final draft of the questionnaire.

Translation

The questionnaire, cover letter, and a postcard followup were prepared in English, Spanish, and French. A professional agency translated all the materials into French and Spanish. Several students at the University of Idaho, whose native language was one of these, wrote critiques about the agency's translations, and numerous inconsistencies were found. The translated documents subsequently were returned to the translation agency and revisions were made. Finally, a foreign-language professor critiqued the revised translations.

Implementation of the Mail Survey

A survey package was airmailed to each park. It consisted of two identical questionnaires about threats, one cover letter explaining the study, a return envelope, and sufficient international reply coupons to pay the cost of return mailing the questionnaire.[2] Managers were requested to keep one questionnaire for themselves and complete and return the other. A postcard was mailed to each park two weeks after the initial mailing, urging managers to complete and return the questionnaire and thanking them if they had already returned it.

Data Coding and Analysis

Responses in each completed questionnaire were numerically coded. Responses written in by non-English-speaking respondents first had to be translated.[3] The data were then keypunched and stored in a disk file so they could be descriptively analyzed with the Statistical Package for the Social Sciences (SPSS; Nie et al. 1975).

The SPSS subprogram FREQUENCIES was used to generate one-way frequency distributions and descriptive statistics for the responses in each questionnaire (Nie et al. 1975). These responses were summarized both cumulatively and by park subsystem. The SPSS subprogram CROSS-TABS was used to generate a chi-square test to determine if the number of respondents who reported a specific threat differed significantly between parks in the different size categories. To minimize statistical error, only parks from two size categories were analyzed.

Responses from parks in each of the four biomes that had the highest response rates were analyzed both cumulatively and by park subsystem. The same statistical tests used to explore differences in responses between parks of different sizes were used to examine responses from parks in the selected biome types. To minimize statistical error, only parks from the two biome types with the highest response rates were analyzed in some tests. Data from parks in countries at different stages of economic development and affiliated and nonaffiliated with special management programs were analyzed in the same manner.

Nonparametric statistics were used for two reasons. The low number of parks and lack of knowledge concerning the distribution of the population required that "distribution-free" tests be used. This was especially important because of the stratified sample. The chi-square was used to test for statistically significant differences between parks along the key independent variables, and lambda was used to measure the strength of the association between independent and dependent variables. For the chi-square analysis, the SPSS program automatically controls for degrees of freedom and provides a significance level. The asymmetric lambda was used because the dependent variable was known in each analysis.

NOTES

1. The biome types for two parks in Jamaica were obtained from IUCN (1982a).

2. International reply coupons usually are accepted in any country " . . . in exchange for postage of that country sufficient to prepay an unregistered surface letter of the first unit of weight (usually 20 grams) to the United States" (U.S. Postal Service 1982:10).

3. Responses written in by respondents were included in the statistical analysis when appropriate.

APPENDIX B

Categorization of Sample by Key Study Variables

Country (Economic Development Stage) and Park	Biome Type No.	Special Management Program? No	Yes	Size (ha)
Australia (more developed)				
1. Croajingolong	6	X		86,000
2. Flinders Ranges	7	X		78,426
3. Hamersley Range	7	X		617,606
4. Kakadu	10		X	614,000
5. Kosciusko	6		X	675,000
6. Lower Glenelg	11	X		27,300
7. South West	2		X	442,240
Bahamas (more developed)				
8. Exuma Cays	13	X		45,584
Benin (developing)				
9. "W"	4	X		502,050
Bolivia (developing)				
10. Ulla Ulla National Faunal Reserve	12	X		137,800
Botswana (less developed)				
11. Chobe	4	X		1,036,000
12. Gemsbok	7	X		2,480,000
Brazil (developing)				
13. Aparados da Serra	2	X		11,307
14. Brasilia	10	X		28,000
15. Iguacu	2	X		170,086
16. Serra de Bocaina	1	X		100,000
Cameroon (less developed)				
17. Waza	4		X	170,000
Canada (more developed)				
18. Banff	12	X		664,076
19. Glacier	12	X		134,939
20. Jasper	12	X		1,087,800
21. Kejimkujik	3	X		38,151
22. Nahanni	3		X	476,560
23. Pacific Rim	2	X		38,850
24. Riding Mountain	3	X		297,591
25. Waterton Lakes	12		X	52,577

continued

Central African Republic (less developed)				
26. Bamingui-Bangoran	4	X		1,070,000
27. Manovo-Gounda-Saint Floris	4	X		1,740,000
Chile (more developed)				
28. Fray Jorge	6		X	9,959
29. Los Paraguas and Conguillio	5	X		58,000
30. Puyehue	2	X		104,017
Colombia (developing)				
31. Cueva de los Guacharos	12		X	9,000
32. Puracé	12		X	83,000
33. Tayrona	4		X	15,000
Congo (developing)				
34. Odzala	1		X	126,600
Costa Rica (developing)				
35. Santa Rosa	4	X		21,500
Czechoslovakia (more developed)				
36. Krkonose	12	X		38,000
Denmark (more developed)				
37. Greenland	9		X	70,000,000
Ecuador (developing)				
38. Galapagos	13		X	691,200
Ethiopia (less developed)				
39. Awash	12	X		72,000
40. Simien Mountains	12		X	22,500
France (more developed)				
41. Cévennes	5	X		84,800
42. Pyrénées occidentales	6	X		47,707
43. Vanoise	12	X		52,839
Gabon (less developed)				
44. Wonga-Wongué	1	X		358,000
Ghana (developing)				
45. Mole	4	X		492,100
Guyana (developing)				
46. Kaieteur	1	X		11,655
Hungary (more developed)				
47. Hortobagy	5		X	52,000
India (developing)				
48. Bandipur	4	X		87,420
49. Kaziranga	4	X		42,994
Indonesia (developing)				
50. Udjung Kulon	1	X		78,619
Italy (more developed)				
51. Abruzzo	12	X		39,160
52. Gran Paradiso	12	X		70,000
Ivory Coast (developing)				
53. Marahoué	1	X		101,000
54. Mont Peko	4	X		34,000
55. Taï	1		X	330,000

continued

Jamaica (developing)				
56. Montego Bay	13	X		1,000
57. Ocho Rios	13	X		1,000
Japan (more developed)				
58. Daisetsuzan	5	X		230,894
59. Nikko	6	X		140,698
Kenya (developing)				
60. Amboseli	7	X		39,206
61. Meru	7	X		87,004
62. Mount Kenya	12	X		71,559
63. Nairobi	4	X		11,721
64. Tsavo	7	X		2,082,114
Malawi (less developed)				
65. Kasungu	4	X		220,300
66. Lengwe	4	X		90,700
67. Nyika	12	X		304,385
Malaysia (developing)				
68. Bako	1	X		2,728
69. Kinabalu	1	X		78,000
Mauritania (less developed)				
70. Banc d'Arguin	7	X		1,200,000
Nepal (less developed)				
71. Royal Chitwan	4	X		93,200
72. Langtang	12	X		170,900
73. Rara	4	X		10,600
74. Sagarmatha	12		X	124,300
Netherlands Antilles (more developed)				
75. Washington	4	X		5,900
New Zealand (more developed)				
76. Fiordland	2	X		1,212,032
77. Mount Cook	2	X		69,957
78. Nelson Lakes	2	X		57,507
79. Urewera	2	X		206,523
Niger (less developed)				
80. "W"	4	X		334,375
Norway (more developed)				
81. Rondane	3	X		57,200
Pakistan (developing)				
82. Kirthar	8	X		308,733
83. Lal suhanra	7		X	31,354
Papua New Guinea (less developed)				
84. McAdam	1	X		2,080
Paraguay (developing)				
85. Ybyku'i	2	X		5,000
86. Cerro Cora	2	X		5,538
Peru (developing)				
87. Manu	1		X	1,532,806
Portugal (developing)				
88. Peneda-Gerês	6	X		60,000

continued

Rwanda (less developed)				
89. Kagera	4	X		251,000
90. Volcanoes	12	X		23,000
Senegal (developing)				
91. Delta du Saloum	4	X		73,000
92. Niokolo-Koba	4		X	913,000
93. Djoudj	4		X	16,000
South Africa (developing)				
94. Bontebok	6	X		2,786
95. Golden Gate Highlands	12	X		4,792
96. Kalahari Gemsbok	7	X		958,103
97. Kruger	4	X		1,948,528
98. Tsitsikama Forest & Coastal	6	X		3,318
Sri Lanka (developing)				
99. Gal Oya	4	X		51,800
100. Wilpattu	4	X		131,884
101. Yala/Ruhuna	1	X		110,000
Sweden (more developed)				
102. Padjelanta	3	X		201,000
103. Peljekaise	3	X		14,600
Switzerland (more developed)				
104. Swiss	12		X	16,887
Tanzania (less developed)				
105. Serengeti	4		X	1,476,300
Thailand (developing)				
106. Khao Yai	1	X		216,800
Tunisia (developing)				
107. Ichkeul	11		X	10,775
Turkey (developing)				
108. Gelibolu Peninsula	7	X		33,000
109. Köprülü Canyon	7	X		36,614
Uganda (less developed)				
110. Ruwenzori	4		X	220,000
USSR (more developed)				
111. Gauya State Reserve	5	X		83,750
United States (more developed)				
112. Big Bend	7		X	286,572
113. Bryce Canyon	8	X		14,405
114. Cape Cod	5	X		18,055
115. Death Valley National Monument	7	X		839,870
116. Everglades	4		X	566,796
117. Grand Canyon	12		X	493,070
118. Great Smoky Mountains	5		X	208,284
119. Mesa Verde	7		X	20,830
120. Olympic	2		X	362,848
121. Padre Island	11	X		54,196
122. Yellowstone	12		X	899,139
123. Yosemite	6	X		308,300
Upper Volta (less developed)				
124. "W"	4	X		190,000

continued

Uruguay (more developed)				
125. Cabo Polonio	11	X		14,250
126. Santa Teresa National Monument	11	X		3,290
Venezuela (more developed)				
127. Canaima	10	X		3,000,000
128. Sierra Nevada	4	X		267,200
129. Yurubi	4	X		23,670
Zaire (developing)				
130. Virunga	12		X	809,000
Zambia (developing)				
131. Kafue	4	X		2,240,000
132. Mosi-Oa-Tunya	4	X		6,600
133. Kasanka	4	X		39,000
Zimbabwe (developing)				
134. Ngezi	4	X		5,800
135. Wankie	4	X		1,465,000

Literature Cited

Abele, L. G., and E. F. Connor. 1979. Application of island biogeography theory to refuge design: Making the right decision for the wrong reasons. Pp. 89–94 in R. M. Linn, ed., *Proceedings of the First Conference on Scientific Research in the National Parks.* Washington, D.C.: U.S. Department of the Interior, Government Printing Office.

Abrahamson, D. 1983. What Africans think about African wildlife. *International Wildlife* 13(4):38–41.

Afolayan, T. A. 1980. A synopsis of wildlife conservation in Nigeria. *Environmental Conservation* 7(3):207–212.

Agee, J. K. 1983. The park experience with natural fire management programs. Presented at the First National Wilderness Workshop, October 11–13, University of Idaho, Moscow, Idaho.

Albright, H. 1918. *1917 annual report of the director of the National Park Service.* 2 vols. Washington, D.C.: National Park Service.

Allen, R. 1980. The world conservation strategy: What it is and what it means for parks. *Parks* 5(2):1–5.

Almond, G. A., and S. Verba. 1963. *The civic culture: Political attitudes and democracy in five nations.* Princeton, N.J.: Princeton University Press.

Armentano, T. V., and O. L. Loucks. 1983. Air pollution threats to U.S. national parks of the Great Lakes region. *Environmental Conservation* 10(4):303–313.

Auerbach, S. I. 1981. Ecosystem response to stress: A review of concepts and approaches. Pp. 29–41 in Barrett, G. W., and R. Rosenberg, eds., *Stress effects on natural ecosystems.* New York: John Wiley and Sons.

Ayensu, E. S. 1984. The afrotropical realm. Pp. 80–86 in McNeely, J. A., and K. R. Miller, eds., *National parks, conservation, and development: The role of protected areas in sustaining society.* Washington, D.C.: Smithsonian Institution Press.

Badshah, M. A. 1962. Parks: Their principles and purposes. In A. Adams, ed., *First World Conference on National Parks.* Washington, D.C.: U.S. Department of the Interior, National Park Service, pp. 24–33.

Baran, P., and E. Hobsbaum. 1961. The stages of economic growth. *Kyklos* 14:234–242.

Barash, D. P. 1982. *Sociobiology and behavior.* 2d ed. New York: Elsevier.

Barnes, R.F.W. 1982. Elephant feeding behavior in Ruaha National Park, Tanzania. *African Journal of Ecology* 20(2):123–136.

______. 1983. Effects of elephant browsing on woodlands in a Tanzanian national park: Measurements, models and management. *Journal of Applied Ecology* 20:521–540.

Barrett, G. W., G. M. Van Dyne, and E. P. Odum. 1976. Stress ecology. *BioScience* 26:192–194.

Barrett, G. W., and R. Rosenberg, eds. 1981. *Stress effects on natural ecosystems.* New York: John Wiley and Sons.

Bayne, B. L. 1975. Aspects of physiological condition in *Mytilus edulis L.* with respect to the effects of oxygen tension and salinity. Pp. 213–238 in Barnes, H., ed., *Proceedings of the ninth European marine biology symposium,* October 2–8, 1974, Oban, Scotland. Aberdeen, Scotland: Aberdeen University Press.

Bekele, E. 1980. *Island biogeography and guidelines for the selection of conservation units for large mammals.* Ph.D. dissertation, University of Michigan, Ann Arbor.

Bennett, J. W. 1976. *The ecological transition: Cultural anthropology and human adaptation.* New York: Pergamon Press.

______. 1984. Ecosystems, environmentalism, resource conservation, and anthropological research. Pp. 289–310 in E. F. Moran, ed., *The ecosystem concept in anthropology.* Boulder, Colo.: Westview Press.

Berry, W. 1981. *The gift of good land.* New York: Brazillier.

Bhagwati, J. 1966. *The economics of underdeveloped countries.* New York: McGraw-Hill.

Blower, J. 1984. Terrestrial parks for developing countries. Pp. 722–727 in McNeely, J. A., and K. R. Miller, eds., *National parks, conservation, and development: The role of protected areas in sustaining society.* Washington, D.C.: Smithsonian Institution Press.

Boardman, R. 1981. *International organization and the conservation of nature.* Bloomington: Indiana University Press.

Boecklen, W. J., and N. J. Gotelli. 1984. Island biographic theory and conservation practice: Species-area or specious-area relationships? *Biological Conservation* 29(1):63–80.

Borg, P. 1977. National park planning and the rights of native peoples. *Parks* 1(4):1–2.

Borman, F. H., and G. E. Likens. 1979. *Pattern and process in a forested ecosystem.* New York: Springer-Verlag.

Borner, M. 1981. Black rhino disaster in Tanzania. *Oryx* 16(1):59–66.

Botkin, D. B., and R. S. Miller. 1974. Complex ecosystems: Models and predictions. *American Scientist* 62(4):448–453.

Bratton, S. P., and P. S. White. 1981. Rare and endangered species management: Potential threats and practical problems in U.S. national parks and preserves. Pp. 459–473 in H. Synge, ed., *The biological aspect of rare plant conservation.* New York: John Wiley & Sons.

Brockman, C. F., and K. Curry-Lindahl. 1962. Committee report—problems of nomenclature: The need for definition. In A. Adams, ed., *First world conference*

on national parks. Washington, D.C.: U.S. Department of the Interior, National Park Service.
Brown, L. R. 1981. *Building a sustainable society.* New York: W. W. Norton and Company.
Brown, L. R., W. U. Chandler, C. Flavin, S. Postel, L. Starke, and E. Wolf. 1984. *State of the world 1984: A Worldwatch Institute report on progress toward a sustainable society.* New York: W. W. Norton and Company.
Bruhn, J. G. 1974. Human ecology: A unifying science? *Human Ecology* 2:105–125.
Budowski, G. 1977. Tourism and conservation: Conflict, coexistence, or symbiosis? *Parks* 1(4):3–6.
Burch, W. R., Jr. 1971. *Daydreams and nightmares: A sociological essay on the American environment.* New York: Harper & Row.
———. 1981. The ecology of metaphor—spacing regularities for humans and other primates in urban and wildland habitats. *Leisure Sciences* 4(3):213–230.
Burch, W. R., Jr., D. DeLuca, G. Machlis, L. Burch-Minakan, and C. Zimmerman. 1978. Handbook for assessing energy-society relations. Report to the U.S. Department of Energy, Office of Inexhaustible Resources, Washington, D.C.
Burch, W. R., and D. DeLuca, eds. 1984. *Measuring the social impact of natural resource policies.* Albuquerque: University of New Mexico Press.
Burgess, J. S., and E. Woolmington. 1981. Threat and stress in the Clarence River Estuary of northern South Wales. *Human Ecology* 9(4):419–431.
Campbell, F. L. 1979. The edge effect: Life in the ecotone. Paper presented at the Second Conference on Scientific Research in the National Parks, San Francisco, Calif.
Caplan, A. L. 1978. *The sociobiology debate.* New York: Harper and Row.
Carpenter, R. A. 1983. *Natural systems for development.* New York: Macmillan.
Catlin, G. 1851. *Illustrations of the manners, customs and conditions of the North American Indians.* 2 vols. London: H. G. Bohn.
Catton, W. R., Jr., and R. E. Dunlap. 1980. A new ecological paradigm for post-exuberant sociology. *American Behavioral Scientist* 24(1):15–47.
Chomsky, N. 1972. *Language and mind.* New York: Harcourt Brace Jovanovich.
Clapham, A. R., ed. 1980. *The IBP survey of conservation sites: An experimental study.* Cambridge: Cambridge University Press.
Clark, C. 1951. *The conditions of economic progress.* 2d ed. London: Macmillan & Co.
Clawson, M. 1973. National parks around the world. *American Forests* 79(3):26–29.
———. 1974. Park visits in the coming decades: Problems and opportunities. Pp. 116–125 in H. Elliott, ed., *Second world conference on national parks.* Morges, Switzerland: International Union for the Conservation of Nature (IUCN).
Clements, F. E. 1916. *Plant succession.* Washington, D.C.: Carnegie Institute.
Commission on National Parks and Protected Areas. 1984. *Threatened protected areas of the world.* Morges, Switzerland: IUCN.

Constantino, I. N. 1974. Present trends in worldwide development of national parks. Pp. 68–80 in H. Elliott, ed., *Second world conference on national parks.* Morges, Switzerland: IUCN.

Cook, E. F. 1976. *Man, energy and society.* San Francisco: W. H. Freeman & Co.

Coolidge, H. J. 1978. Evolution of the concept, role and early history of national parks. Pp. 29–33 in R. Osten, ed., *World national parks: Progress and opportunities.* Brussels: IUCN.

Cottrell, W. F. 1951. Death by dieselization: A case study in the reaction to technological change. *American Sociological Review* 16(3):358–365.

Curry-Lindahl, K. 1972. Ecological research and management. Pp. 197–213 in R. Osten, ed., *World national parks: Progress and opportunities.* Brussels: IUCN.

______. 1974a. The conservation story in Africa during the 1960s. *Biological Conservation* 6(3):170–177.

______. 1974b. The global role of national parks for the world of tomorrow. The Horace M. Albright Conservation Lectureship no. 14, May 23. Berkeley: University of California, School of Forestry and Conservation.

Darling, F. F. 1969. Wilderness, science and human ecology. Pp. 198–214 in W. Schwartz, ed., *Voices for the wilderness.* New York: Ballantine Books.

Darling, F. F., and N. D. Eichhorn. 1967. *Man and nature in the national parks—reflections on policy.* Washington, D.C.: Conservation Foundation.

Dasmann, R. F. 1973. A system for defining and classifying natural regions for purposes of conservation. IUCN Occasional Paper no. 7. Morges, Switzerland: IUCN.

______. 1974. Biotic provinces of the world. IUCN Occasional Paper no. 9. Morges, Switzerland: IUCN.

______. 1975. National parks, nature conservation, and future primitive. Pp. 89–97 in *Proceedings of the South Pacific conference on national parks and reserves,* February 24–27, Wellington, New Zealand.

______. 1984. The relationship between protected areas and indigenous peoples. Pp. 667–671 in McNeely, J. A., and K. R. Miller, eds., *National parks, conservation, and development: The role of protected areas in sustaining society.* Washington, D.C.: Smithsonian Institution Press.

Dasmann, R. F., J. P. Milton, and P. H. Freeman. 1973. *Ecological principles for economic development.* New York: John Wiley and Sons.

Davies, N. B. 1978. Ecological questions about territorial behavior. Pp. 317–350 in Krebs, J. R., and N. B. Davies, eds., *Behavioural ecology: An evolutionary approach.* Oxford: Blackwell.

DeBellevue, E., H. T. Odum, J. Browder, and G. Gardner. 1979. Energy analysis of the Everglades National Park. Pp. 31–43 in *Proceedings of the first conference on scientific research in the national parks.* Washington, D.C.: U.S. Department of the Interior, National Park Service.

deFreitas, C., and E. Woolmington. 1980. Catastrophe theory and catastasis. *Area* 2(3):191–194.

deGroot, R. S. 1983. Tourism and conservation in the Galapagos Islands. *Biological Conservation* 26:291–300.

De Santo, R. S. 1978. *Concepts of applied ecology.* New York: Springer-Verlag.

Devall, B. 1980. The deep ecology movement. *Natural Resources Journal* 20 (Spring):299–322.

Diamond, J. M. 1975. The island dilemma: Lessons of modern biogeographic studies for the design of natural reserves. *Biological Conservation* 7(2):129–146.

di Castri, F., and J. Robertson. 1982. The biosphere reserve concept: 10 years after. *Parks* 6(4):1–6.

Dice, L. R. 1952. *Natural communities.* Ann Arbor: University of Michigan Press.

Dickenson, R. E. 1984. Keynote address: The Nearctic realm. Pp. 492–495 in McNeely, J. A., and K. R. Miller, eds., *National parks, conservation, and development: The role of protected areas in sustaining society.* Washington, D.C.: Smithsonian Institution Press.

Dillman, D. A. 1978. *Mail and telephone surveys—the total design method.* New York: John Wiley & Sons.

Dubos, R. 1980. *The wooing of earth.* New York: Charles Scribner's Sons.

Duncan, O. D. 1964. Social organization and the ecosystem. Pp. 36-82 in F. Robert, ed., *Handbook of modern sociology.* New York: Rand McNally.

Dunlap, R. E., and W. R. Catton. 1983. What environmental sociologists have in common (whether concerned with "built" or "natural" environment). *Sociological Inquiry* 53(2-3):113–135.

Eckholm, E. 1982. *Down to earth: A report on the environment and human needs.* New York: W. W. Norton & Co.

Eicher, C. K., and J. M. Staatz. 1983. *Agricultural development in the third world.* Baltimore: Johns Hopkins University Press.

Eidsvik, H. K. 1984. Future directions for the Nearctic realm. Pp. 546–549 in McNeely, J. A., and K. R. Miller, eds., *National parks, conservation, and development: The role of protected areas in sustaining society.* Washington, D.C.: Smithsonian Institution Press.

Eisinger, R. A., W. P. Janicki, R. L. Stevenson, and W. L. Thompson. 1974. Increasing returns in international mail surveys. *Public Opinion Quarterly* 38(1):124–130.

Ellen, R. 1982. *Environment, subsistence and system: The ecology of small-scale social transformations.* Cambridge: Cambridge University Press.

Elliott, H., ed. 1974. *Second world conference on national parks.* Morges, Switzerland: IUCN.

Eltringham, S. K., and R. C. Malpas. 1980. The decline in elephant numbers in Rwenzori and Kabelega Falls National Parks, Uganda. *African Journal of Ecology* 18(1):73–86.

Eltringham, S. K., and M. H. Woodford. 1973. The numbers and distribution of buffalo in the Rwenzori National Park, Uganda. *East African Wildlife Journal* 11(2):141–150.

Emlen, J. M. 1973. *Ecology: An evolutionary approach.* Reading, Mass.: Addison-Wesley.

Erdos, P. L., and A. J. Morgan. 1970. *Professional mail surveys.* New York: McGraw-Hill.

Food and Agriculture Organization, United Nations. 1982. *Tropical forest resources.* Forestry Paper no. 30. Rome, Italy.

Forster, R. 1973. Planning for man and nature in national parks. IUCN Publication no. 26. Gland, Switzerland: IUCN.

Frome, M. 1981. What is happening to our national parks? *National Parks* 55:10–15.

Gardner, J. E., and J. G. Nelson. 1981. National parks and native peoples in Northern Canada, Alaska and Northern Australia. *Environmental Conservation* 8(3):207–215.

Garratt, K. 1984. The relationship between adjacent lands and protected areas: Issues of concern for the protected area manager. Pp. 65–71 in McNeely, J. A., and K. R. Miller, eds., *National parks, conservation, and development: The role of protected areas in sustaining society.* Washington, D.C.: Smithsonian Institution Press.

Garrett, W. E. 1978. The Grand Canyon: Are we loving it to death? *National Geographic* 154(1):16–51.

Gilbert, F. S. 1980. The equilibrium theory of island biogeography: Fact or fiction? *Journal of Biogeography* 7:209–235.

Gilbert, L. E. 1980. Food web organization and its conservation of neotropical diversity. Pp. 11–33 in Soulé, M. E., and B. A. Wilson, eds., *Conservation biology: An evolutionary-biological perspective.* Sunderland, Mass.: Sinauer Associates.

Goddard, M. K. 1961. What the United States can learn. Pp. 35–43 in H. Jarrett, ed., *Comparisons in resource management—six notable programs in other countries and their possible U.S. application.* Baltimore: Johns Hopkins University Press.

Gomm, R. 1974. The elephant man. *Ecologist* 4(2):53–57.

Gorio, S. 1978. Papua New Guinea involves its people in national park development. *Parks* 3(2):12–14.

Hales, D. F. 1984. The world heritage convention—status and directions. Pp. 744–750 in McNeely, J. A., and K. R. Miller, eds., *National parks, conservation, and development: The role of protected areas in sustaining society.* Washington, D.C.: Smithsonian Institution Press.

Halfon, E. 1975. The systems identification problem and the development of ecosystem models. *Simulation* 25(6):149–152.

Harris, L. D. 1984. *The fragmented forest: Island biogeography theory and the preservation of biotic diversity.* Chicago: University of Chicago Press.

Harris, M. 1980. *Cultural materialism.* New York: Random House.

Harrison, J., K. Miller, and J. A. McNeely. 1982. The world coverage of protected areas: Development goals and environmental needs. *Ambio* 11(5):238–245.

Harroy, J. P. 1974. A century in the growth of the national park concept throughout the world. Pp. 24–32 in H. Elliott, ed., *Second world conference on national parks.* Morges, Switzerland: IUCN.

Hart, W. J. 1966. *A systems approach to park planning.* Morges, Switzerland: IUCN.

Hawley, A. H. 1950. *Human ecology: A theory of community structure.* New York: Ronald Press.

Hill, M. 1983. Kakadu National Park and the aboriginals: Partners in protection. *Ambio* 12(3-4):158–169.

Hillman, K., and E. Martin. 1979. Will poaching exterminate Kenya's rhinos? *Oryx* 15(2):131–132.

Hinrichsen, D., ed. 1983. How the world heritage convention works. *Ambio* 12(3-4):140–145.

Holdgate, M. W., M. Kassas, and G. F. White, eds. 1982. *The world environment 1972–1982: A report by the United Nations Environment Programme.* Natural resources and the environment series vol. 8. Dublin, Ireland: Tycooly International Publishing.

Hughes, A. J. 1979. Myths of the tourist industry. *Africa Report* 24(3):39–43.

International Union for the Conservation of Nature (IUCN). 1975. *World directory of national parks and other protected areas.* Gland, Switzerland: IUCN.

———. 1982a. List of biosphere reserves. *IUCN Bulletin* 13(7-8-9):69.

———. 1982b. *United Nations list of national parks and equivalent reserves.* Gland, Switzerland: IUCN.

———. 1984. Categories, objectives and criteria for protected areas. Pp. 47–53 in McNeely, J. A., and K. R. Miller, eds., *National parks, conservation, and development: The role of protected areas in sustaining society.* Washington, D.C.: Smithsonian Institution Press.

Ise, J. 1961. *Our national park policy—a critical history.* Baltimore: Johns Hopkins University Press.

Jarvie, I. C. 1963. Theories of cargo bults: A cultural analysis. *Oceania* 34:1.

Jefferies, B. E. 1982. Sagarmatha National Park: The impact of tourism in the Himalayas. *Ambio* 11(5):247–281.

Jellicoe, G., and S. Jellicoe. 1975. *The landscape of man.* London: Thomas & Hudson.

Johnson, J. B., M. W. Young, and C. O. French. 1980. Coastal zone studies: A holistic approach. Paper presented at the Wildlife Management Institute, North American Wildlife and Natural Resource Fifty-fourth Conference, March 22–26, Miami Beach, Florida.

Kitching, G. 1982. *Development and underdevelopment in historical perspective.* London: Methuen & Co.

Kitching, R. L. 1983. *Systems ecology: An introduction to ecological modeling.* St. Lucia, Australia: University of Queensland Press.

Klausner, S. F. 1971. *On man and his environment.* San Francisco: Jossey-Bass.

Krebs, C. J. 1972. *Ecology: The experimental analysis of distribution and abundance.* New York: Harper & Row.

———. 1978. Optimal foraging: Decision rules for predators. Pp. 23–63 in Krebs, J. R., and N. B. Davies, eds., *Behavioural ecology: An evolutionary approach.* Oxford: Blackwell.

Kusel, J. 1984. Vitalism, mechanism and the emergence of the systems paradigm in contemporary science. Unpublished report, Cooperative Park Studies Unit. Moscow: University of Idaho.

Kusler, J. A. 1974. Public/private parks and management of private lands for park protection. IES Report no. 16. Madison: University of Wisconsin–Madison, Institute for Environmental Studies.

Kwapena, N. 1984. Wildlife management by the people. Pp. 315–321 in McNeely, J. A., and K. R. Miller, eds., *National parks, conservation, and development:*

The role of protected areas in sustaining society. Washington, D.C.: Smithsonian Institution Press.

Lamprey, H. P. 1974. Management of flora and fauna in national parks. Pp. 237–248 in H. Elliott, ed., *Second world conference on national parks.* Morges, Switzerland: IUCN.

Lazlo, E. 1972. *Introduction to systems philosophy.* New York: Gordon and Breach Science Publishers.

Lee, R. F. 1968. *Public use of the national park system 1872–2000.* Washington, D.C.: U.S. Department of Interior, National Park Service.

Lemmons, J., and D. Stout. 1982. National parks legislative mandate in the United States of America. *Environmental Management* 6(3):199–207.

Livingston, J. 1982. *The fallacy of wildlife conservation.* Toronto: McLelland and Steward.

Lucas, P.H.C. 1982. Major issues of the future: How protected areas can help meet society's evolving needs. Paper presented at the World National Parks Congress, October 11–22, Bali, Indonesia.

Lumsden, C. J., and E. O. Wilson. 1981. *Genes, mind, and culture: The coevolutionary process.* Cambridge: Harvard University Press.

Lusigi, W. J. 1981. New approaches to wildlife conservation in Kenya. *Ambio* 109(2-3):87–92.

McCloskey, M. 1984. World parks. *Sierra* 69(6):36–42.

McKenzie, R. D. 1926. The scope of human ecology. *American Journal of Sociology* 32:1.

McNeely, J. A., and K. R. Miller. 1984. *National parks, conservation, and development: The role of protected areas in sustaining society.* Washington, D.C.: Smithsonian Institution Press.

MacArthur, R. H., and E. O. Wilson. 1967. *The theory of island biogeography.* Princeton, N.J.: Princeton University Press.

Machlis, G. E. 1984. Protection forestry. In *The human factors affecting forestry/fuelwood projects: An agenda for research and development.* U.S. AID Workshop, February 11–13, 1984, Washington, D.C.

Machlis, G. E., and W. R. Burch, Jr. 1983. Relations between strangers: Cycles of structure and meaning in tourist systems. *Sociological Review* 31(4):665–692.

Machlis, G. E., and R. G. Wright. 1984. A method for surveying the state of the parks. Cooperative Park Studies Unit Report CPSU/UI SB84-4. Moscow: University of Idaho.

Machlis, G. E., D. R. Field, and F. L. Campbell. 1981. The human ecology of parks. *Leisure Sciences* 4(3):195–212.

Machlis, G. E., and D. R. Field, eds. 1984. *On interpretation: Sociology for interpreters of natural and cultural history.* Corvallis: Oregon State University Press.

Maragos, J. E., A. Soegiarto, E. D. Gomez, and M. A. Dow. 1983. Development planning for tropical coastal ecosystems. Pp. 229–298 in R. A. Carpenter, *Natural systems for development.* New York: Macmillan.

Marks, S. A. 1984. *The imperial lion: Human dimensions of wildlife management in central Africa.* Boulder, Colo.: Westview Press.

Martindale, D. 1960. *American society.* New York: D. Van Nostrand.
Meier, G. M., and R. E. Baldwin. 1961. *Economic development: Theory, history, policy.* New York: Wiley and Sons.
Micklin, M. 1977. The ecological perspective in the social sciences: A comparative overview. *Leisure Research* 1(4):53–68.
Miller, J. G. 1978. *Living systems.* New York: McGraw-Hill.
Miller, K. R. 1984. The natural protected areas of the world. Pp. 20–23 in McNeely, J. A., and K. R. Miller, eds., *National parks, conservation, and development: The role of protected areas in sustaining society.* Washington, D.C.: Smithsonian Institution Press.
Miller, R. I. 1978. Applying island biogeographic theory to an East African reserve. *Environmental Conservation* 5:191–195.
Mishra, H. R. 1982. Balancing human needs and conservation in Nepal's Royal Chitwan Park. *Ambio* 11(5):246–251.
Moran, E. F. 1984a. Limitations and advances in ecosystems research. Pp. 3–32 in E. F. Moran, ed., *The ecosystem concept in anthropology.* Boulder, Colo.: Westview Press.
Moran, E. F., ed. 1984b. *The ecosystem concept in anthropology.* Boulder, Colo.: Westview Press.
Mueller-Dombois, D., K. Kartawinata, and L. L. Handley. 1983. Conservation of species and habitats: A major responsibility in development planning. Pp. 1–51 in R. A. Carpenter, ed., *Natural systems for development: What planners need to know.* New York: Macmillan.
Mugangu-Trinto, E. 1983. A new approach for the management of Zairean national parks. Cooperative Park Studies Unit Report CPSU/UI B83-3. Moscow: University of Idaho.
Murdoch, W. W. 1980. *The poverty of nations: The political economy of hunger and population.* Baltimore: Johns Hopkins University Press.
Musoke, M. B. 1980. Overbrowsing of *Capparis tomentosa* by goats in Rwenzori National Park, Uganda. *African Journal of Ecology* 18(1):7–10.
Myers, N. 1972. National parks in savannah Africa. *Science* 178(4067):1255–1263.
______. 1981. A farewell to Africa. *National Wildlife* 11(6):36–46.
______. 1984. Too big for their own good? Sheer size puts five of Africa's largest mammals in direct conflict with people. *International Wildlife* 14(3):26–32.
Myers, N., and D. Myers. 1982. From the "duck pond" to the global commons: Increasing awareness of the supranational nature of emerging environmental issues. *Ambio* 11(4):195–201.
Naess, A. 1973. The shallow and the deep, long-range ecology movements: A summary. *Inquiry* (Oslo) 16:95–100.
______. 1984. Identification as a source of deep ecological attitudes. Pp. 256–270 in M. Tobias, ed., *Deep ecology.* San Diego, Calif.: Avant Books.
National Parks and Conservation Association (NPCA). 1979. NPCA adjacent lands survey: No park is an island. *National Parks and Conservation Magazine* 53(3):4–9.

Nelson, J. G. 1978. International experience with national parks and related reserves. Pp. 1–27 in J. G. Nelson, R. D. Needham, and D. L. Mann, eds., *International experience with national parks and related reserves.* University of Waterloo, Ontario: Department of Geography.

Nelson, J. G., R. D. Needham, and D. L. Mann, eds. 1978. *International experience with national parks and related reserves.* University of Waterloo, Ontario: Department of Geography.

Nicholson, M. 1974. What is wrong with the national park movement? Pp. 32–37 in H. Elliott, ed., *Second world conference on national parks.* Morges, Switzerland: IUCN.

Nie, N., C. H. Hull, J. G. Jenkins, K. Steinbrenner, and D. H. Bent. 1975. *Statistical package for the social sciences.* 2d ed. New York: McGraw-Hill.

Odum, E. P. 1971. *Fundamentals of ecology.* 3d ed. Philadelphia: W. B. Saunders.

Odum, E. P., J. T. Finn, and E. H. Franz. 1979. Perturbation theory and the subsidy-stress gradient. *BioScience* 27:349–352.

Odum, H. T. 1971. *Environment, power and society.* New York: Wiley Interscience.

______ . 1983. *Systems ecology: An introduction.* New York: Wiley and Sons.

Olwig, K. F. 1980. National parks, tourism and local development: A West Indian case. *Human Organization* 39(1):22–30.

Ovington, J. D. 1984. Ecological processes and national park management. Pp. 60–64 in McNeely, J. A., and K. R. Miller, eds., *National parks, conservation, and development: The role of protected areas in sustaining society.* Washington, D.C.: Smithsonian Institution Press.

Park, R. E. 1936. Human ecology. *American Journal of Sociology* 42(1):1–15.

Pianka, E. R. 1983. *Evolutionary ecology.* 3d ed. New York: Harper and Row.

Polunin, N., and H. K. Eidsvik. 1979. Ecological principles for the establishment and management of national parks and equivalent reserves. *Science* 155:1203–1207.

Rappaport, R. A. 1967. *Pigs for the ancestors: Ritual in the ecology of a New Guinea people.* New Haven, Conn.: Yale University Press.

Rapport, D. J., and H. A. Regier. 1980. An ecological approach to environmental information. *Ambio* 9(1):22–27.

Redfield, R. 1963. *The little community/Peasant society and culture.* Chicago: University of Chicago Press.

Ricklefs, R. E. 1973. *Ecology.* Portland, Or.: Chiron Press.

Risser, G. P., and K. D. Cornelison. 1979. *Man and the biosphere.* Norman: University of Oklahoma Press.

Robertson, I. 1981. *Sociology.* 2d ed. New York: Worth Publishers.

Roe, F. G. 1955. *The Indian and the horse.* Norman: University of Oklahoma Press.

Rosa, E. A., and G. E. Machlis. 1983. Energetic theories of society: An evaluative review. *Sociological Inquiry* 53(2/3):152–178.

Rostow, W. W. 1971. *The stages of economic growth.* 2d ed. New York: Cambridge University Press.

Runte, A. 1979. *National parks: The American experience.* Lincoln: University of Nebraska Press.

Sahlins, M. 1976. *The use and abuse of biology.* Ann Arbor: University of Michigan Press.

Savidge, J. M. 1968. Elephants in the Ruaha National Park, Tanzania—management problem. *East African Agriculture and Forestry Journal* 33:191–196.

Sax, J. L. 1980a. Buying scenery: Land acquisitions for the National Park Service. *Duke Law Journal* 1980(4):709–740.

______. 1980b. *Mountains without handrails—reflections on the national parks.* Ann Arbor: University of Michigan Press.

______. 1982. The compromise called for. In E. H. Connally, ed., *National parks in crisis.* Washington, D.C.: National Parks and Recreation Association.

Scavia, D., and A. Robertson, eds. 1979. *Perspectives on lake ecosystem modelling.* Ann Arbor, Mich.: Ann Arbor Science.

Schonewald-Cox, C. M. 1983. Conclusions: Guidelines to management: A beginning attempt. Pp. 414–445 in Schonewald-Cox, C. M., S. M. Chambers, B. MacBryde, and W. L. Thomas, eds., *Genetics and conservation—a reference for managing wild animal and plant populations.* Menlo Park, Calif.: Benjamin/Cummings Publishing.

Selye, H. 1952. *The story of the adaptation syndrome.* Montreal: Acta.

Shands, W. E. 1979. *Federal resource lands and their neighbors.* Washington, D.C.: Conservation Foundation.

Shelford, V. E. 1963. *The ecology of North America.* Urbana: University of Illinois Press.

Shelton, N. 1983. Parks and sustainable development—1982 World National Parks Congress. *National Parks and Conservation* 57(5-6):16–21.

Sicinski, A. 1970. "Don't know" answers in cross-national surveys. *Public Opinion Quarterly* 34(1):126–129.

Siehl, G. H. 1971. *Visitor pressures on national parks.* Washington, D.C.: Congressional Research Service, Environmental Policy Division.

Simberloff, D. S., and L. G. Abele. 1982. Refuge design and island biogeography theory: Effects of fragmentation. *American Naturalist* 120:41–50.

Slayter, R. 1984. The world heritage convention: Introductory comments. P. 734 in McNeely, J. A., and K. R. Miller, eds., *National parks, conservation, and development: The role of protected areas in sustaining society.* Washington, D.C.: Smithsonian Institution Press.

Smith, E. A. 1984. Anthropology, evolutionary ecology, and the explanatory limitations of the ecosystem concept. Pp. 51–85 in E. F. Moran, ed., *The ecosystem concept in anthropology.* Boulder: Westview Press.

Soulé, M. E., B. A. Wilcox, and C. Holtby. 1979. Benign neglect: A model of faunal collapse in the game reserves of East Africa. *Biological Conservation* 15(4):259–260.

State of California. 1983. Stewardship—1983: Managing the natural and scenic resources of the California State Park System. Unpublished report. Sacramento: California Department of Parks and Recreation.

Stearns, S. C. 1977. The evolution of life history traits: A critique of the theory and a review of the data. *Annual Review of Ecology and Systematics* 8:145–171.

Steward, J. H. 1955. The concept and method of cultural ecology. In J. Steward, *The theory of culture change.* Urbana: University of Illinois Press.

Stottlemyer, J. R. 1981. Evolution of management policy and research in the national parks. *Journal of Forestry* 79(1):16–20.

Suttles, G. D. 1972. *The social construction of communities.* Chicago: University of Chicago Press.

Theodorson, G. A., and A. Theodorson. 1969. *Modern dictionary of sociology.* New York: Thomas Y. Crowell.

Tichnell, D. L., and G. E. Machlis. 1984. Threats to national parks: An international survey. Cooperative Park Studies Unit Report CPSU/UI S84-1. Moscow: University of Idaho.

Tichnell, D. L., G. E. Machlis, and J. R. Fazio. 1983. Threats to national parks: A preliminary survey. *Parks* 8(1):14–17.

Tobey, G. B. 1973. *A history of landscape architecture—the relationship of people to the environment.* New York: American Elsevier.

Tobias, M., ed. 1984. *Deep ecology.* San Diego, Calif.: Avant Books.

Todaro, M. P. 1981. *Economic development in the third world.* 2d ed. New York: Longman.

Turnbull, C. 1972. *The mountain people.* New York: Simon and Schuster.

Udvardy, M.D.F. 1975. A classification of the biogeographical provinces of the world. IUCN Occasional Paper no. 18. Morges, Switzerland: IUCN.

———. 1984. A biogeographical classification system for terrestrial environments. Pp. 34–38 in McNeely, J. A., and K. R. Miller, eds., *National parks, conservation, and development: The role of protected areas in sustaining society.* Washington, D.C.: Smithsonian Institution Press.

United Nations. 1979. World population trends and prospects, 1950–2000. New York: United Nations.

United Nations Educational, Scientific, and Cultural Organization (UNESCO), United Nations Environment Programme, Food and Agricultural Organization. 1978. *Tropical forest ecosystems: A state-of-knowledge report.* Paris: UNESCO.

U.S. Congress, House, Conservation, Energy, Natural Resource Subcommittee. 1976. *Degradation of the national parks.* 94th Congress, 1st and 2d sess. Washington, D.C.

U.S. Council on Environmental Quality. 1980. *The global 2000—report to the president.* 3 vols. New York: Penguin Books.

U.S. Department of the Interior (USDI), National Park Service. 1980. *State of the parks 1980—a report to Congress.* Washington, D.C.

U.S. Postal Service. 1982. International postal rates and fees. Publication no. 51. Washington, D.C.: Government Printing Office.

Vayda, A. P., and R. A. Rappaport. 1976. Ecology, cultural and noncultural. In Richerson, P. J., and J. McEvoy III, eds., *Human ecology: An environmental approach.* North Scituate, Mass.: Duxbury Press.

Von Bertalanffy, L. 1956. *General systems.* New York: Brazillier.

———. 1968. *General systems theory.* New York: Brazillier.

von Droste zu Hülshoff, B. 1984. How UNESCO's Man and the Biosphere Programme is contributing to human welfare. Pp. 689–691 in McNeely, J.

A., and K. R. Miller, eds., *National parks, conservation, and development: The role of protected areas in sustaining society.* Washington, D.C.: Smithsonian Institution Press.

Wagner, R. H. 1971. *Environment and man.* New York: W. R. Norton.

Warwick, D. P., and S. Osherson. 1973. Comparative analysis in the social sciences. Pp. 3–41 in *Comparative research methods.* Englewood Cliffs, N.J.: Prentice-Hall.

Wauer, R. H., and W. R. Supernaugh. 1983. Wildlife management in the national parks—a historical perspective. *National Parks* 57(7/8):12–16.

Western, D. 1982. Amboseli National Park: Enlisting landowners to conserve migratory wildlife. *Ambio* 11(5):302–308.

Western, D., and J. Ssemakula. 1981. The future of savannah ecosystems: Ecological islands or faunal enclaves? *African Journal of Ecology* 19:7–19.

Wetterberg, G. B. 1974. The history and status of South American national parks and an evaluation of selected management options. Ph.D. dissertation, University of Washington.

______. 1984. The exchange of wildlands technology: A management agency perspective. Pp. 728–732 in McNeeley, J. A., and K. R. Miller, eds., *National parks, conservation, and development: The role of protected areas in sustaining society.* Washington, D.C.: Smithsonian Institution Press.

White, L., Jr. 1967. The historical roots of our ecological crisis. *Science* 155:1203–1207.

White, P. S., and S. P. Bratton. 1980. After preservation: Philosophical and practical problems of change. *Biological Conservation* 18(1980):241–255.

Wielgolaski, F. E. 1971. National parks and other protected areas in North America in relation to those in Norway and Sweden. *Biological Conservation* 3(4):285–291.

Wilcox, B. A. 1980. Insular ecology and conservation. Pp. 95–117 in M. E. Soule', and B. A. Wilcox, eds., *Conservation biology: An evolutionary-ecological perspective.* Sunderland, Mass.: Sinauer.

Willson, M. F., and S. W. Carothers. 1979. Avifauna of habitat islands in the Grand Canyon. *Southwestern Naturalist* 24:563–576.

Wilson, E. O. 1975. *Sociobiology: The new synthesis.* Cambridge: Harvard University Press.

______. 1978. *On human nature.* Cambridge: Harvard University Press.

Wolf, R. 1982. Crisis in the national parks. *Rocky Mountain Magazine* 4(11):49–55.

Wolff, W. J. 1982. Selection and management of coastal and estuarine protected areas. Presented at the World National Parks Congress, October 11–22, Bali, Indonesia.

World Bank. 1981. *World development report.* New York: Oxford University Press.

Wright, G. M., J. S. Dixon, and B. H. Thompson. 1933. Fauna of the national parks of the United States. U.S. Department of the Interior, National Park Service, Fauna Series no. 1, Washington, D.C.

Wright, R. G., and G. E. Machlis. 1984. Models for park management: A prospectus. Cooperative Park Studies Unit Report CPSU/UI SB85-1. Moscow: University of Idaho.

Yoaciel, S. M. 1981. Changes in the populations of large herbivores and in the vegetation community in Mweya Peninsula, Rwenzori National Park, Uganda. *African Journal of Ecology* 19:303–312.

Young, O. 1982. *Resource regimes: Natural resources and social institutions.* Berkeley: University of California Press.

Index